D1171810

New Perspectives on Evolution

THE WISTAR SYMPOSIUM SERIES

The Wistar
Symposium Series
Volume 4

New Perspectives on Evolution

Proceedings of a Multidisciplinary Symposium Designed to
Interrelate Recent Discoveries and New Insights
in the Field of Evolution
Held at the University of Pennsylvania,
April 18 and 19, 1990

Sponsored by
The Wistar Institute
Philadelphia, Pennsylvania

Edited by

Leonard Warren
Hilary Koprowski

A JOHN WILEY & SONS, INC., PUBLICATION
New York • Chichester • Brisbane • Toronto • Singapore

Address all Inquiries to the Publisher
Wiley-Liss, Inc., 41 East 11th Street, New York, NY 10003

Printed in United States of America

Recognizing the importance of preserving what has been written, it is a policy of John Wiley & Sons, Inc. to have books of enduring value published in the United States printed on acid-free paper, and we exert our best efforts to that end.

Library of Congress Cataloging-in-Publication Data

New perspectives on evolution: proceedings of a multidisciplinary symposium designed to interrelate recent discoveries and new insights in the field of evolution held at the University of Pennsylvania, April 18 and 19, 1990/sponsored by the Wistar Institute, Philadelphia, Pennsylvania; editors, Leonard Warren, Hilary Koprowski.
p. cm.—(The Wistar symposium series; v. 4)
Includes bibliographical references and index.
ISBN 0-471-56068-5
1. Evolution—Congresses. I. Warren, Leonard, 1924– . II. Koprowski, Hilary. III. Wistar Institute of Anatomy and Biology. IV. Series.
QH359.N48 1991 90-45462
575—dc20 CIP

Contents

Contributors

Mark B. Adams, Departments of History and Sociology of Science, University of Pennsylvania, Philadelphia, PA 19104 **[37]**

Garland E. Allen, Department of Biology, Washington University, St. Louis MO 63130; present address: Department of the History of Science, Harvard University, Cambridge, MA 02138 **[15]**

Raoul E. Benveniste, Laboratory of Viral Carcinogenesis, National Cancer Institute, Frederick Cancer Research and Development Center, Frederick, MD 21701-1013 **[225]**

Mitchell Bush, National Zoological Park, Smithsonian Institution, Washington, DC 20008 **[225]**

Rebecca L. Cann, Department of Genetics, University of Hawaii at Manoa, Honolulu, HI 96822 **[209]**

Russell F. Doolittle, Center for Molecular Genetics, University of California, San Diego, La Jolla, CA 92093 **[165]**

Mary A. Eichelberger, Laboratory of Viral Carcinogenesis, National Cancer Institute, Frederick Cancer Research and Development Center, Frederick, MD 21701-1013 **[225]**

Walter Gilbert, Department of Cellular and Developmental Biology, Harvard University, Cambridge, MA 02138 **[155]**

David Goldman, Laboratory of Clinical Studies, National Institute of Alcohol Abuse and Alcoholism, National Institutes of Health, Bethesda, MD 20892 **[225]**

Daniel L. Hartl, Department of Genetics, Washington University School of Medicine, St. Louis, MO 63110 **[123]**

Jonathan J. Henry, Institute for Molecular and Cellular Biology, Department of Biology, Indiana University, Bloomington, IN 47405 **[189]**

Margaret G. Kidwell, Department of Ecology and Evolutionary Biology, University of Arizona, Tucson, AZ 85721 **[139]**

Andrew H. Knoll, The Botanical Museum, Harvard University, Cambridge, MA 02138 **[77]**

Janice S. Martenson, Laboratory of Viral Carcinogenesis, National Cancer Institute, Frederick Cancer Research and Development Center, Frederick, MD 21701-1013 **[225]**

Ernst Mayr, Museum of Comparative Zoology, Harvard University, Cambridge, MA 02138 **[1]**

William G. Nash, H & W Cytogenetic Services, Inc., Lovettsville, VA 22080 **[225]**

Stephen J. O'Brien, Laboratory of Viral Carcinogenesis, National Cancer Institute, Frederick Cancer Research and Development Center, Frederick, MD 21701-1013 **[225]**

Kenneth R. Peterson, Department of Ecology and Evolutionary, Biology, University of Arizona, Tucson, AZ 85721 **[139]**

Rudolf A. Raff, Institute for Molecular and Cellular Biology, Department of Biology, Indiana University, Bloomington, IN 47405 **[189]**

Mitchell L. Sogin, Center for Molecular Evolution, Marine Biological Laboratories, Woods Hole, Falmouth, MA 02543 **[175]**

Steven M. Stanley, Department of Earth and Planetary Sciences, Johns Hopkins University, Baltimore, MD 21218; present address: Department of Geological Sciences, Case Western Reserve University, Cleveland, OH 44106 **[87]**

Keith Stewart Thomson, The Academy of Natural Sciences, Philadelphia, PA 19103 **[101]**

Nikolai N. Voronstov, N.K. Kol'tsov Institute of Developmental Biology, USSR Academy of Sciences, 11734 Moscow, USSR **[65]**

Robert K. Wayne, Department of Biology, University of California, Los Angeles, Los Angeles, CA 90024 **[225]**

David E. Wildt, National Zoological Park, Smithsonian Institution, Washington, DC 20008 **[225]**

Gregory A. Wray, Institute for Molecular and Cellular Biology, Department of Biology, Indiana University, Bloomington, IN 47405 **[189]**

Foreword

The Wistar Institute's 1966 monograph on evolution, *Mathematical Challenges to the Neo-Darwinian Interpretation of Evolution,* included contributions by Ernst Mayr, Hilary Koprowski, Martin Kaplan, Sir Peter Medawar, Loren Eiseley, Richard Lewontin, C. H. Waddington, Murray Eden, and Sewall Wright, among others. A kind of pervasive skepticism and argumentation runs through the entire volume; at that time there were many prominent mathematicians and physical scientists who felt that biologists could not understand how the evolutionary process worked—that it could be described only in physical and mathematical terms.

Since 1966, man has explored the moon and found it uninteresting. The most gigantic telescopes aimed at other planets have failed to detect signs of life. Thus, thanks to fate or to the organizer(s) of the universe, we need not bother particularly about evolution in relation to celestial systems other than earth. On the other hand, the giant strides made in the molecular approaches to gene function, particularly in the analysis and manipulation of DNA, permit the study of interspecies relationships on a much more factual basis than ever dreamt by Darwin.

The papers in this volume discuss the astonishing advances that have been made in our knowledge and understanding of evolution since the 1960s. Rather than to select one or a few aspects of evolution for discussion, our approach has been panoramic, albeit at the expense of depth in any one area. However, some emphasis is placed on the spectacular contributions to evolutionary knowledge made by molecular biology, contributions that in 1966—and even at the start of the last decade—were quite unimaginable. Research methods of remarkable sensitivity and analytic power are now almost commonplace, and these are providing so much information that without the computer, the unaided human brain would be overwhelmed.

This book will not answer all—perhaps not even many—of the questions concerning evolution. We are still and always will be searching for better concepts and ideas. Montaigne pondered the question and wrote:

> But I am dissatisfied with my mind in that it usually brings forth its profoundest ideas, as well as its maddest and those I like best,

> unexpectedly, and when I least look for them, for they will instantly vanish if I have no means at hand for fixing them; on horseback, at table, in bed, but mostly on horseback, where my thoughts wander most widely. (*The Essays,* translated by E. J. Trechmann, London 1935 [2]:336).

Shall we ride to our next symposium on horseback?

Hilary Koprowski
Director
The Wistar Institute
Philadelphia, Pennsylvania

Preface

New Perspectives in Evolution is the result of a special symposium held at The Wistar Institute in 1989, one of a series sponsored by Wistar on matters of broad interest to the scientific community as well as to the Institute's own biomedical researchers. Evolution was especially appealing as a topic for one of these conferences, linked as it is to developmental biology and genetics, two areas of research that have been pursued at The Wistar Institute since its founding in 1892.

"New Perspectives" was actually the second Wistar symposium with an evolutionary theme. In 1966 "Mathematical Challenges to the Neo-Darwinian Interpretation of Evolution" was held at Wistar and recorded as an influential monograph under the editorship of Paul S. Moorhead and Martin M. Kaplan. Originally published by the Wistar Institute Press, demand for this volume was such that it was reprinted by Alan R. Liss, Inc. in 1985. The present collection of articles, covering current evolutionary thought and theory from the diverse perspectives of a number of the field's most respected arbiters, promises to meet with a similar reception—and, we hope, comparable longevity as food for thought.

Even the recent advances in our understanding of the evolutionary process could not be reviewed comprehensively in the limited space of a single publication. However, from a world of possibilities, an illustrious group of experts have contributed papers that are interesting and enlightening and that will give readers a deeper and more rounded appreciation of this central, mammoth, inescapable process of living matter.

We feel especially fortunate that the venerable Ernst Mayr, one of the great formulators of our present notions about evolution and an eminent contributor to the 1966 publication, opened the 1989 symposium and wrote an overview of the field to provide historical perspective for this volume.

In addition, it was fortuitous that Nikolai Vorontsov—who was recently appointed to Gorbachev's cabinet as minister of the environment—was in Philadelphia at the time of the symposium. Epic changes are taking place in the Soviet Union, a country where political and social philosophy dominated, perverted, and almost destroyed the study of evolution, where Swiftian Yahoos once dominated any rational approach to genetics. Dr.

Vorontsov's paper is a moving testimony as witness to the age of darkness in Russia in the middle third of our century.

We thank Matthew Meselson of Harvard University for his help in organizing the symposium and for serving as chairman of one of the sessions.

It is a pleasure to express our appreciation to Zoe Zampana, Monica Winter, Diana Burgwyn, Marie Lennon, Cherise Kent, Colleen Cannon, and Betty Warthen, who worked so effectively for the success of the symposium. Bonnie Clause was an organizing force in the publication of the proceedings.

Finally, we gratefully acknowledge the generous financial support provided by the Alfred P. Sloan Foundation, the Dolphinger-McMahon Foundation, and Wiley-Liss, Inc.

Leonard Warren
Institute Professor
The Wistar Institute
Philadelphia, Pennsylvania

New Perspectives on Evolution, pages 1-14

Introduction: An Overview of Current Evolutionary Biology

ERNST MAYR
Alexander Agassiz Professor, Emeritus, Museum of Comparative Zoology, Harvard University, Cambridge, Massachusetts 02138

INTRODUCTION

A scientist in an active field, as is evolutionary biology, is like an explorer in a mountainous country: every new ridge he climbs opens up new vistas and previously hidden terrain. Every new discovery poses new challenges.

Evolutionary biology, in the 130 years since Darwin, experienced a series of such major conquests that revealed new peaks to be conquered. Such conquests were Weismann's demonstration of the constancy of the genetic material, Mendel's reaffirmation of the particulateness of the genetic material, and finally, the evolutionary synthesis, which resulted in the general acceptance of natural selection, contributed the explanation of organic diversity, and reemphasized the populational nature of evolutionary phenomena. Since Gar Allen will deal with the earlier history, I will limit myself to a discussion of the evolutionary synthesis, and post-synthesis developments.

What was the synthesis? It was not an intellectual revolution, but a unification of the numerous strands of evolutionary biology. It must be remembered that much progress in science consists of the refutation of erroneous or at least invalid theories. This is abundantly true for the evolutionary synthesis and, historically considered, this is perhaps its major significance. The synthesis ratified the definitive refutation of the three leading anti-Darwinian theories, saltationism, Lamarckism, and orthogenesis (i.e., a teleological view of evolution). What the architects of the synthesis proposed instead, was not a finished, well-rounded theory, hence it is somewhat misleading to speak of *the* synthetic theory. Rather, the synthesis resulted in a research program based on sound Darwinian principles, yet continuously in need of further elaboration and modification in minor points. As I hope to show in the ensuing review, the Darwinism of the evolutionary synthesis has been so successful during the last 40 years in answering any and all criticisms, that the expectation that it might have to be replaced by a new evolutionary theory, would seem to be very ill-founded. Almost all recent criticisms of the theory of the synthesis were based on ignorance, hence it is important to repeat again and again what the basic insights are, that were supplied by the synthesis. What are these basic new insights? Perhaps the greatest contribution made by the evolutionary synthesis was that it signalled the beginning of a new way of thinking. It consisted of the refutation of all deterministic tendencies, which had entered evolutionary biology from two sources, from the teleological thinking so prevalent in the pre-Darwinian period (and retained by all the orthogenetic theories), and from physicalist thinking. Beginning with the synthesis, and increasingly so after it, such determinism was replaced by a recognition of the frequency of stochastic processes and of pluralism, that is, of multiple solutions to problems posed by nature. The other major achievement of the synthesis was a far-reaching clarification of concepts, leading to an elimination of previously prevailing confusion.

I shall now take up the major subjects of evolutionary biology, and discuss in what respects the evolutionary synthesis brought a clarification, also in what areas there are still open frontiers.

EVOLUTION

The naturalists who contributed so much to the evolutionary synthesis showed how incomplete if not misleading was the reductionist definition of evolution, as a change in gene frequencies. As I have pointed out previously (Mayr 1977, 1982) this definition quite misses the point. Organic evolution is described far better as "a change in adaptation and in biological diversity." Changes in gene frequencies are merely a byproduct of these more basic processes. Furthermore, it is questionable to what extent changes in the frequency of neutral genes can be designated as evolution. The now rejected definition is most nearly correct for prokaryotes, but it is singularly inappropriate for complex higher organisms.

But this is not all! Even today it is not realized by some of those who write about evolution that there are three entirely different, in fact incompatible, concepts of evolution (Mayr 1988:457).

1. *Saltational evolution* refers to changes that are due to the sudden production of new types. This is the only kind of evolution conceivable for an essentialist.
2. *Transformational evolution.* According to this concept, clearly articulated by Lamarck, evolution consists of the gradual transformation of a thing from one condition of existence to another. This is the only kind of evolution found in inanimate nature, for instance, in cosmology, and it is this kind of evolution which the developmental biologists study when they study the ontogeny of an individual from the egg stage to adulthood. Significantly, in German, the same word Entwicklung is used both for organic evolution and for ontogeny. It is the erroneous transfer

of this concept of transformational evolution from ontogeny to organic evolution that has been largely responsible for the long-standing opposition of developmental biologists to Darwinism.

3. *Variational evolution.* As Lewontin (1983) has pointed out, Darwin introduced an entirely new concept of evolution. According to it, new "gene pools" (as we would now say) are generated in every generation, and evolution takes place because the successful individuals produced by these gene pools give rise to the next generation. Thus, an entirely new start is, so to speak, made in every generation. Evolution is thus merely contingent on certain processes articulated by Darwin: Variation and selection. Such evolution is not necessarily progressive; it does not strive toward perfection or toward any other goal; it is opportunistic, hence unpredictable.

Darwinian evolution is not a unitary process, but consists of two types of processes: The origin of diversity and the origin of adaptation. In this symposium I will not say much about the evolution of diversity, it relates to the multiplication of species, species selection, punctuation, and some aspects of macroevolution. If one wants to study the origin of diversity, one must adopt multi-dimensional thinking and population thinking. An interest in the origin of diversity was virtually absent among the mathematical population geneticists and was brought into the synthesis by the naturalists and those geneticists who, as Chetverikov and Dobzhansky, had been raised as naturalists.

The prevalence of adaptation was for the natural theologians the most conspicuous aspect of living nature. They could explain it only by postulating design by an intelligent creator. It was Darwin's enormous achievement to have been able to explain adaptation materialistically, that is without invoking supernatural forces. He said that adaptation can only be explained as the result of natural selection, even though it has been realized, particularly since the evolutionary synthesis, that all sorts of chance phenomena and stochastic processes contribute to evolutionary change.

NATURAL SELECTION

Darwin's basic insight was the incredible fertility of most organisms, who produce hundreds, thousands or even millions of offspring, while on an average only two among this multitude are necessary to perpetuate the population. There is no finalistic component in Darwin's model that would drive toward perfection; the two survivors are not in any teleological manner selected in order to survive. The process of selection is, in principle, an optimization process, but owing to its probabilistic nature, to its constraints, and to the frequency of stochastic processes, it can not achieve optimality.

Darwin never understood the nature and the origin of genetic variability. He simply accepted its universal availability as a black box. After the elucidation of the problems of variability by genetics, the production of variability has become for us an important component of the process of natural selection. In fact, it greatly facilitates the understanding of natural selection if one says that it is a two-step process, the first step consisting of the production of genetic variability, and the second step dealing with the differential survival of these variants, selection in the narrow sense.

The two-step concept of natural selection also helps to clarify the relationship between adaptation and selection. Some modern authors have objected to the formulation that all adaptation is due to natural selection. What they really objected to was a teleological concept of natural selection as a process driving toward adaptation. I am sure we all reject such a teleological interpretation. Any variation produced during the first step of natural selection that is not preadapted, is not likely to be selected for during the second step. Hence every variant, no matter what its adaptive potential is, has to pass through the sieve of the second step in order to be permanently incorporated in the genome of the population. This is how the formula "all adaptation is due to natural selection" must be understood.

Selection throughout is highly opportunistic. It is, as Francois Jacob has said so rightly, a process of tinkering. In order to construct anything new, it makes use of any structure or behavior or other component of variability that happens to be available. This is beautifully illustrated by the diversity of plumage characters that are employed by male birds of paradise to enhance their showiness; or, to give a second example, by the great multitude of mechanisms and features used by pelagic animals to keep floating in the water. Owing to this opportunism of selection it is quite unpredictable what features will evolve next.

The nature of selection is made even clearer when one makes a distinction between "selection of" and "selection for," as perceptively recommended by Sober (1984). The question *selection of,* or in other words, what is the target of selection, is in almost all cases answered by an individual. Such an individual might be a single cell, as in the case of prokaryotes and protists, or a complex whole organism, as in the case of higher animals and plants. Saying that the individual is the principal target of selection does not negate the possibility of additional selection at higher hierarchical levels, such as when whole species compete with each other. Indeed the species-specific characteristics of species A may be such that they give every individual of this species superiority over any individual of species B. This superiority may lead to the extinction of species B. Such species selection is effected by individual selection, and is not in conflict with it. Darwin excellently described such species selection that had resulted from the introduction of British animals and plants into New Zealand.

The question of *selection for* may be answered by a single gene. The case of the sickle cell allele of human hemoglobin is a typical case, and so are, in fact, the genes of many other human hereditary diseases. Nevertheless, as I pointed out recently (Mayr 1988), I do not know of a single case where a single gene rather than an individual is the target of selection. The major reason why so many geneticists adopted the gene as the target of selection was that it facilitated their calculations. Their erroneous formulation ignored the fact that the selective value of a gene may depend quite considerably on its genetic background (the total genotype).

To accept the individual rather than the gene as the target of selection has the additional advantage that it is consistent with the insight that recombination is of far greater significance than mutation in the evolutionary change from generation to generation. It is genetic recombination, at least in higher organisms, which produces the material with which natural selection works. The new individuals produced in every generation by new recombination are the target of selection; they are the real material of natural selection. Knowing that the organism as a whole is the target of selection focuses attention on the considerable difference between single-celled prokaryotes and highly complex multicellular higher organisms. In the latter, owing to their slow generation time, it would be totally improbable that the right mutation would always occur to cope with a newly arisen challenge from the environment. It is recombination that gives them the needed genetic variability and flexibility.

The discovery by Jukes, Kimura, and others of the high frequency of neutral or quasi-neutral base pair replacements, caused considerable consternation among those who defined evolution as a change in gene frequencies." "How could selection favor neutral mutations?," they asked. This puzzle is solved as soon as one realizes that the individual as a whole—and its genotype as a whole—is the real target of selection. As long as the individual as a whole is favored by selection, it does not matter how many neutral or even slightly deleterious changes it may carry along as hitchhikers. There is no mystery in neutral evolution, nor is it in any conflict with Darwinian theory.

The new insight, however, poses a considerably more difficult puzzle. It is the question of the organization of the genotype. I will come back to this point when I discuss molecular biology. At this time I will only mention that the importance of epistatic interactions in the genotype becomes obvious as soon as one adopts the concept of the individual as a whole as the target of selection. Furthermore, it has now become evident that we can no longer treat all genes and gene components as if they were identical in evolutionary importance. Exons, introns, flanking sequences, promoters, regulatory genes, transposons, and whatever other kinds of genes or gene components there are, all may differ from each other in evolutionary significance. We can no longer base all of our conclusions on one single kind of such genes, let us say enzyme genes as studied by electrophoresis.

To me the study of the respective evolutionary roles of these different categories of genes is one of the great open frontiers of evolutionary biology. I am sure that an understanding of the differences of the selective significance of various kinds of genes and gene components will help to answer many of the open questions of this field. There are many empirical findings, particularly in the study of macroevolution, as for instance the extraordinary stability if not inertness of certain evolutionary lineages, for which we have no convincing explanation so far. A study of the role of different kinds of genes may perhaps tell us why drastic genetic reorganizations during peripatric speciation are possible when such events of rapid and drastic reconstruction are apparently rare if not impossible under other conditions. To be sure these phenomena are of no great relevance in the study of single gene pools or for the minor adaptive changes that occur in a succession of populations in the time dimension. However, these problems are of major importance in the solving of the problems of macroevolution, as I shall presently discuss.

TWO KINDS OF SELECTION

Actually, as has become very clear in the last 15 years, there are two kinds of selection; perhaps one could also say there are two kinds of properties on which there is a selective premium. They were already recognized and distinguished by Darwin under the names natural selection and sexual selection. Natural selection refers to anything that favors survival, whether it is an increase or decrease of body size, a broadening of the niche or its more efficient utilization, better protection against the environment or an increased tolerance of environmental extremes, a superior ability to cope with diseases or to escape enemies, that is, anything that would accomplish greater ecological-physiological efficiency or save energy. Any individual favored by such selection would contribute genotypes to the gene pool that are apt to spread in future generations and thus enhance the adaptedness of the population as a whole.

However, Darwin also saw—much more clearly than any of his contemporaries—that not all selection leads to improved adaptedness. An individual might contribute more genes to the next generation not through physiological efficiency or any other component of viability, but simply through being more successful in producing offspring. This kind of selection Darwin called sexual selection.

The phenomenon of sexual selection was neglected by evolutionists for 100 years even though Darwin had devoted almost two thirds of the text of his *The Descent of Man* (1871) to this subject. At the present moment it is perhaps this component of selection that is studied most intensively. Selection for reproductive success deals not only with competition among males, or the choice of optimal males by females, but with all sorts of other phenomena, such as generational conflict, sib rivalry, and numerous other phenomena investigated in the contemporary behavioral-ecological literature. The current intense interest in selection for reproductive success, and in the behaviors and devices that tend to favor such a success, is undoubtedly one of the few major developments in post-synthesis evolutionary biology (Trivers 1985). At one time I referred to selection for reproductive success "as a weakness in the armor" of natural selection (Mayr 1963:199). Up to a point this is indeed true, because such selection very often reduces the overall survival propensity of those individuals (and their offspring) that are favored by this selection. We do not know whether it has ever contributed to the decline and possibly even the extinction of certain species, but this is quite possible.

VARIATION

No doubt we will hear much in this symposium on the production of genetic variation, and its availability for natural selection. Perhaps I should here mention two misconceptions among the opponents of Darwinism. The first is the belief that variation can produce anything. Actually, variation is extremely constrained within a given taxon. Cuvier already remarked that one would never find a carnivore with horns. Weismann emphasized again and again how limited the kind of variation is that is available to a given group of organisms. Geneticists have abundantly proven how strong these constraints are. No Drosophila will ever be found with blue, bright yellow, or vividly green eyes. There are multiple kinds of constraints, which I have recently discussed in detail (Mayr 1988: 106–113).

There is a second aspect of variation that is often misrepresented in the literature. When a Darwinian states that mutations are random, he does not mean in the slightest that any mutation is possible. All he means is, first, that the locus of the next mutation is not predictable, and second, that a mutation is neither an answer to the needs of an organism nor the result of a response to a particular set of environmental conditions. Actually, as I have already said, the number and type of mutations possible in a particular organism are extremely limited. It is possible that my unrestricted claim of randomness will be refuted in the near future, at least for single-celled organisms, since I am fully aware of the recent researches of John Cairns. However, as of this moment, so many different interpretations of his findings have been given that it does not seem advisable to change the traditional Darwinian definition of randomness in evolution prematurely. However, such a change may indeed be necessary in the future.

I have stated that natural selection can only deal with the variation that is offered to it. There are now cases known where genetic variation is biased to such an extent that it can override the effectiveness of natural selection. The segregation distorter genes are such a case. Most biased variation may be due to the action of retroviral-like transposons. Some cases of biased variation have been cited in the literature as cases of gene selection. This is a misinterpretation, because even in these cases, the variation comes first, and only then is the resulting zygote or reproducing cell exposed to natural selection.

DARWINISM AND DEVELOPMENT

A number of recent critics of Darwinism have accused the architects of the evolutionary synthesis of having neglected the importance of development. I am afraid these critics have things upside down. It was not the architects of the synthesis who neglected development, but the students of development who ignored the synthesis. This is obvious when one looks at the history of the evolutionary synthesis. After Dobzhansky in 1937 had outlined the synthesis in bold strokes, specialists in various other biological disciplines joined him by showing not only that their specialty also fitted in with the concepts of the synthesis but also by showing that their particular fields could make some very specific contributions to the synthesis. This was true for Mayr in 1942 (for systematics), for Simpson in 1944 (for paleontology), for Rensch in 1947 (also 1939, 1943) (for general zoology), and for G. L. Stebbins in 1950 (for botany). Developmental biology would have been most welcome to join. This would have been an important event, because developmental biology, owing to the success of Entwicklungsmechanik, was perhaps the most flourishing branch of biology in the 1920s and 1930s. However, most embryologists were quite hostile to Darwinism (Waddington was an exception). Their thinking was controlled by the concept of transformational evolution, in fact I suspect that virtually all the embryologists of the period more or less adhered to Lamarckism (see also Hamburger 1980). The concept of variational evolution was utterly alien to them. It is quite characteristic for the current criticism of Darwinism that most of these critics come from the tradition of developmental biology. They are still sure that Darwinism is wrong, and they look for a new paradigm (see for instance Ho and Saunders 1984, Pollard 1984, and Reid 1985). I have commented on the extraordinary ignorance of the authors of two of these books in recent reviews (Mayr 1984, 1986) and so has Burian (1988). In spite of all the efforts of these authors, they have been quite unable to discover even a single phenomenon of developmental biology that is incompatible with the theories of the evolutionary synthesis. These authors fail to make a distinction between evolutionary causations (origins of new genetic programs) and proximate causations (phenomena of the translation of these genetic programs into phenotypes). When they invoke Aristotle they seem to be unaware of the extraordinary similarity of Aristotle's eidos and the genetic program (Delbrück 1971; Gotthelf 1976). From Darwin on, evolutionists have been interested in development, particularly since it offers so many puzzling problems. Every stage in the life cycle of an organism from the moment of the fertilization of the egg until death is a potential target of natural selection. But this is entirely in line with the thinking of Darwinism.

To be sure, we still have no explanation for certain phenomena in ontogeny. Our fish-like ancestors came on land about 400 million years ago. They shifted to a mode of respiration that no longer required gills. Yet all land-living vertebrates from amphibians to birds and mammals still go through a gill arch stage in their ontogeny. Why are the structures of the neck region not developed directly instead of in the roundabout way of first forming gill arches? On a purely theoretical basis a Darwinian would say these embryonic gill arches are retained in ontogeny because they must have a selective significance. What could this significance be?

Explanatory endeavors based on the traditional concepts and terminologies of genetics and developmental biology have not been very illuminating. However, by using the language of information science one can articulate an explanation that would seem consistent with the empirical evidence. Let us begin with the question: What controls the construction of an organ? In the case of all structures and organs that are controlled very directly by the action of certain genes, we say that construction is controlled by the genetic program. However, in behavior for instance, what the genetic program does is to create certain structures in the central nervous system which then functions as the program that later in life controls behavioral activities. One could call such secondary programs somatic programs, in contradistinction to the genetic programs in the original blueprint of development. Every embryologist knows how many steps in development are caused, or at least strongly influenced, by somatic structures or tissues that had previously developed. These structures or tissues can be considered somatic programs, which, together with the genetic program, control the subsequent origin or growth of other structures. Accepting this line of thought one can suggest that the embryonic gill arches function as somatic programs assisting in the development of the structures of the neck region, analogously to Spemann's organizer and other similar theoretical constructs of developmental biology. I am sure every developmental biologist will to a large extent agree with this interpretation. The main point, however, I want to make is that such an explanation is in every respect compatible with the Darwinian paradigm.

I think it will help the discussion if I take up this problem once more with a different approach.

MOLECULAR BIOLOGY AND EVOLUTION

Early in the history of molecular biology there was a widespread feeling that its new discoveries might necessitate a complete rewriting of evolutionary theory. Indeed some molecular biologists did claim that their findings had refuted some of the tenets of Darwinism. Most of these claims were based on findings of biased genetic variation. Undeniably, many findings of molecular biology are in conflict with the picture of classical genetics. None perhaps was as startling as the discovery that genes are highly complex systems consisting of exons, introns, flanking sequences, and all that. But have any of these discoveries required a revision of Darwinism? I claim that they have not. Let me document this by listing some of the findings of molecular biology that are most important for the understanding of evolution.

1. The genetic program does not by itself supply the building material of new organisms, but only the blueprint for the making of the phenotype.
2. The pathway from nucleic acids to proteins is a one-way street. Proteins and any information they may have acquired during their lifetime are not translated back into nucleic acids (the so-called central dogma).
3. Not only the genetic code but in fact most basic molecular mechanisms are the same in all organisms from the most primitive prokaryotes up (Mayr 1988:538–539).

Another finding of molecular genetics, although up to now, it cannot be formulated in very precise terms, is that the genotype is a highly complex system that should not be looked at simply as an aggregate of genes with additive action. This conclusion, at the present time, is perhaps more of an article of faith than a solid piece of knowledge. Non-additive interactions, of course, are difficult to experiment with, considering that every genotype is unique. They are difficult to study and refractory to analysis. However, there are numerous evolutionary phenomena that suggest the existence of such interacting properties of the genotype. I have discussed them in numerous previous publications, and will have to leave my assertions at that.

MACROEVOLUTION

And this brings us to the question of macroevolution and development. The question is often asked, how can the origin of new types, of entirely new structures, and of new higher taxa, be explained in terms of selective advantages of individuals that are parts of a population. The mentioned three phenomena are generally bracketed together as macroevolution. At first sight, there seems to be no connection between the changes of gene frequencies within a local population and the phenomena of macroevolution. Indeed, as Ayala (1983) has rightly stated, microevolutionary phenomena within a single population and the phenomena of macroevolution seem to be decoupled. And yet it is equally obvious that the phenomena of macroevolution must be controlled by the genotypes of individuals. How can these seemingly contradictory viewpoints be reconciled? It is at this point that we have to introduce hierarchical thinking. Between the individual and the higher taxa are two other hierarchical levels, those of the population and of the species. The origin of species, which was such a puzzle to the essentialists, was solved by showing that isolated populations could serve as the intermediate steps in the process of speciation.

If I were to be asked what I consider to be the most important current challenge in evolutionary biology, I would say that it is the elucidation of the structure of the genotype.

Mathematical population genetics introduced a highly reductionist approach to evolutionary questions by defining evolution as a change in gene frequencies and by considering the gene to be the unit or target of selection. The existence of non-additive aspects of the genotype was mentioned by them in discussions, but was never seriously taken in consideration in their explanatory theories.

I protested as early as 1950 against this gene-centered theorizing, which I maligned as bean-bag genetics. Subsequently Sewall Wright chided me for having included him among the bean-bag geneticists, and he was justified, because indeed he had called attention to epistatic interactions in his discussions, even though most of his graphs and mathematics dealt with the frequency of single genes. Such a concentration on the frequency of single genes was very productive in the history of population genetics, but it has limits. The actual study of genic interactions has made little progress in the last 30 years.

Why do we want to know more about the structure of the genotype? The reason is that there are many evolutionary phenomena that are inconsistent with an atomistic, strictly additive view of the genotype.

Several speakers on our program will discuss evolutionary phenomena relating to this problem. This includes the evolutionary inertia of most mature species, it includes the failure of any new animal phyla to have originated in the 500 million years since the Cambrian in contrast to the explosive evolution of 70 new Bauplans in a less than 100 million year period after the origin of the metazoans, it includes the rapidity of evolution that oft accompanies speciation in founder populations, and various other evolutionary phenomena.

All these puzzling questions can be answered, at least in part, if one assumes that there is, what I have called "a cohesion of the genotype." By this I mean that the genes are not independent of each other, but have a tendency to join in teams or systems, that control certain developmental processes. Furthermore, it appears that throughout evolution there has been a selective premium for making such intergenic connections in the course of time ever tighter, so that eventually the whole genotype, or at least part of it, "congeals," as somebody has called it. When this happens, the major problem of evolution becomes how to loosen up such tightly constrained genotypes. My proposal of genetic revolutions in founder populations (Mayr 1954, 1963) was, I believe, the first recognition of this problem.

I am convinced that a collaboration between molecular and developmental biologists will eventually solve the puzzling problem how such constrained domains of the genotype can continue to exist in the face of crossing over and recombination.

I want to add one historical comment. Some of the recent critics have attacked the synthesis for being atomistic and reductionist. This is an error.

It was only the school of the mathematical geneticists that can be accused of this attitude. The accusation, however, is not applicable to those architects of the synthesis such as myself and Rensch, for whom the individual as a whole always has been the target of selection, as it had been for Darwin. Rejecting the Fisherian atomism was actually a return to a more genuine Darwinism, notwithstanding the contrary claims by certain critics.

IS THE EVOLUTIONARY SYNTHESIS STILL ALIVE?

The question is justified after the frequent attacks on the evolutionary synthesis in recent years, whether the synthesis is still a valid set of theories, or whether it should be replaced by a new and better paradigm. It is obvious from my discussion that I am of the opinion that nothing is seriously wrong with the achievements of the evolutionary synthesis and that it does not need to be replaced. What is most encouraging for a Darwinian is that none of the recent critics has been able to propose a sound alternative to Darwinism. Not a single new theory has been proposed in the last 50 years that can be considered a serious challenge to the paradigm developed during the evolutionary synthesis. To be sure, our understanding of evolution concerning certain details is still incomplete. We still do not understand why the genotype is seemingly so stable in certain evolutionary lines and so easily modified in others. We are still somewhat uncertain what happens during certain modes of speciation. We still do not know to what an extent biased variation can override the force of selection. We are still rather in the dark as to the nature and persistence of the manifold epistatic interactions within a genotype, to mention only a few of our areas of ignorance. However, in none of these questions do I see any possible solution that would be in conflict with the basic Darwinian paradigm. What we must learn, if we have not learned it already, is that the physicalist approach of reductionism, of determinism, and of universal laws, is rarely suitable for evolutionary explanations. Outcomes of natural selection are probabilistic and there are often several alternate possible solutions of selection pressures for novel adaptations. We have to have a regard for the historical component in the genetic program of every single organism. We must at all times realize that we have to solve questions concerning both the evolutionary and the proximate causations, that stochastic processes are responsible for a good deal of evolutionary noise, and that evolution, as a historical process, is not controlled by universal laws so characteristic of the physical sciences (Jacob 1983). At the same time evolutionists have found nothing at the cellular or the molecular level that is in any conflict whatsoever with the laws of physics and chemistry. Darwinian evolution has liberated us from the non-materialistic schemes for the explanation of the living world. Whatever open questions are left, and whatever controversies an evolutionary biologist may have to face in the future, I am confident that they can all be resolved within the solid framework of Darwinism.

CONCLUSIONS

I'm quite sure that each of the problems that I have mentioned will turn up again and again in the ensuing presentations, and that will be the time to answer my arguments if someone would like to do so. I would just like to say that we have all realized, more so in recent years than earlier, that the particular constellation of ideas at any given period—what in German is called the Zeitgeist—controls to a large extent what scientific theories will be successful and what others will not, and what particular theories will even be advanced and which will not. The reason that, Darwinism, for instance, from 1859 to the evolutionary synthesis, let's say 1940, never could become universally adopted not even by the biologists, was clearly due to the fact that there was a whole set of philosophical ideas which were incompatible with natural selection. They gradually had to be made obsolete, downbeaten, and so forth before this could happen. There was a whole series of such philosophical ideas and Garland Allen will now tell us about this part of the subject, speaking on "Philosophical Crossroads and Controversies in 20th Century Evolutionary Theory."

REFERENCES

Ayala FJ (1983): Beyond Darwinism? The challenge of macroevolution to the synthetic theory of evolution. East Lansing, PA: Phil Sci Assn 1982 2:275–291.

Burian RM (1988): Challenge to the evolutionary synthesis. Evol Biol 23:247–269.

Delbrück M (1971): Aristotle-totle-totle. In Monod J, Borek E (eds): "Of Microbes and Life." New York: Columbia Univ Press.

Gotthelf A (1976): Aristotle's conception of final causality. Rev Metaphysics 30:226–254.

Hamburger V (1980): Evolutionary theory in Germany: A comment. In Mayr E, Province WB (eds): "The Evolutionary Synthesis," Cambridge (MA): Harvard Univ Press, pp 303–308.

Ho MW, Saunders PT (eds) (1984): "Beyond Neo-Darwinism: An Introduction to the New Evolutionary Paradigm." New York: Academic Press.

Jacob F (1977): Evolution and tinkering. Science 196:1161–1166.

Lewontin R (1983): The organism as the subject and object of evolution. Scientia 118:63–82.

Mayr E (1954): Change of genetic environment and evolution. In Huxley J (ed): "Evolution as a Process." London: Allen and Unwin, pp 157–180.

Mayr E (1963): "Animal Species and Evolution." Cambridge (MA): Harvard Univ Press.

Mayr E (1977): The study of evolution historically viewed. In Goulden CE (ed): "The Changing Scenes in Natural Sciences 1776–1976." Philadelphia: Acad Nat Sci, Spec Publ 12, pp 39–58.

Mayr E (1982): "The Growth of Biological Thought." Cambridge (MA): Harvard Univ Press.

Mayr E (1984): The triumph of the evolutionary synthesis. Times Lit Suppl 257(4):1261–1262.

Mayr E (1986): [Review of] Reid RGB (1985), "Evolutionary Theory: The Unfinished Synthesis." Isis 77:358–359.

Mayr E (1988): "Toward a New Philosophy of Biology." Cambridge (MA): Harvard Univ Press.

Pollard JW (ed) (1984): "Evolutionary Theory: Paths into the Future." New York: John Wiley and Sons.

Reid RGB (1985): "Evolutionary Theory: The Unfinished Synthesis." Ithaca, NY: Cornell Univ Press.

Sober E (1984): "The Nature of Selection: Evolutionary Theory in Philosophical Focus." Cambridge (MA): MIT Press.

Trivers R (1985): "Social Evolution." Menlo Park, CA: Benjamin/Cummings.

A NOTE ON *TINKERING*

In discussions of evolution the term *tinkering* has sometimes received a whipping and been used in a rather derogatory way. Under these circumstances it might be worthwhile to remember who introduced the term, and why he introduced it.

The person who introduced tinkering was Francois Jacob of the Institut Pasteur. In continental Europe, and in France in particular, there is an incredibly strong tradition—far stronger than in the United States—of thinking that everything has to be leading to a progressive end, that everything is teleological, finalistic, or—even if you reject that kind of philosophical approach—at least physicalist. The thinking was very strong that everything is due to deterministic laws, and that natural selection is part of this stream and that it inevitably, one might say predictably, leads to certain ends. This thinking was what Francois Jacob was fighting. He reminds us that Jacques Monod was fighting exactly the same thing when he published his book on necessity or chance, because necessity was in all the theories that both Monod and Jacob encountered in college.

The important thing is that, while natural selection makes use of the second step, as I have called it, in the process of natural selection, the first step is entirely controlled at random, admittedly very strongly constrained. There is nothing lawlike there, there's nothing teleological, there's nothing predictive. And now transferring it to the second step, this step—being natural selection in the more narrow sense of the word—makes use of whatever it encounters that had been produced by the first step. And that is exactly as Jacob says, it works as a *tinkerer* who uses odd pieces of wires or nails, or pieces of wood and so forth, to put together what he wants to assemble, without any longterm planning going into what he does.

Thus the words *tinkerer* and *tinkering* are merely ways of describing an emphasis on the absence of the deterministic and finalistic, thus teleological, or anything of that sort in the process of natural selection. The word *tinkering* should be treated with a certain amount of respect, rather than trying to make it ridiculous.

Ernst Mayr

New Perspectives on Evolution, pages 15-36

Mechanistic and Dialectical Materialism in 20th Century Evolutionary Theory: The Work of Ivan I. Schmalhausen

GARLAND E. ALLEN
Department of Biology, Washington University, St. Louis, Missouri 63130

INTRODUCTION

> An upsurge of activity unprecedented since the time immediately following the publication of Darwin's 'Origin of Species' has taken place in the last ten to fifteen years in the field of evolutionary biology.

So wrote Theodosius Dobzhansky in his forward to Ivan Schmalhausen's *Factors of Evolution* in 1949 (Schmalhausen 1949; p. ix). Although Dobzhansky was referring to the period we now call the classical "evolutionary synthesis," he could well have been speaking (prophetically) of the present. There is today much ferment and controversy about many components of traditional Darwinian theory and we are expanding our horizons about the process of evolution—now to the molecular level. Such expansions in thought inevitably evoke controversy and dissent. In this paper I would like to discuss one philosophical dimension of controversy in evolutionary theory that has persisted from past to present.

There have, in fact, been three major periods of controversy within the field of evolutionary theory in the present century. The first occurred between 1890 and 1910, when Darwinian theory was under attack by biologists of all sorts: Embryologists, systematicists, geneticists, and paleontologists. Darwinism was, in the words of one of its critics, "on its deathbed" (Dennert 1903). A second period of ferment was that to which Dobzhansky referred, the classical synthesis of Darwinian theory with Mendelian genetics, systematics and paleontology, between 1920 and 1940. This period was characterized not only by the excitement attending development of a new field of integrated study (population genetics), but also by a series of debates on such issues as the most effective population size for selection, the amount of variation within natural populations (Provine 1971; Mayr and Provine 1980) and the role of isolation in producing divergence. The third period is the present—or the past decade, at any rate—when we find ourselves in the midst of controversies about punctuated change, the extent and significance of extra, redundant, or "garbage" DNA within the genome, the effects of drift, the extent of mass extinctions, and whether microevolutionary processes are sufficient to explain macroevolutionary change.

Although each of these periods of controversy has been marked by its own particular issues, a persistent theme has underpinned many of these discussions ever since Darwin. This theme revolves around a continuing debate between two very different schools of philosophical thought. On the one hand, from the early through the middle 19th century the dominant philosophical view of mechanistic materialism provided a particularly productive way of viewing phenomena in all areas of natural science—most specifically, physics, and chemistry. It was thus not unusual to find the mechanistic viewpoint, with its emphasis on breaking complex problems down into their component parts, quantification, experimentation and rigorous analysis, being applied in the development of Darwinian theory as well as to other areas of biology such as embryology and genetics.

At the same time, the mechanistic viewpoint, with its analogy of organisms and all natural processes to machines, with its atomism and attendant reductionism, seemed too simplistic to a number of biologists as a method for understanding the large-scale evolutionary process. In the latter part of the 19th through the early decades of the 20th century physiologists such as Walter Bradford Cannon and Lawrence J. Henderson, embryologists such as F. R. Lillie, Hans Spemann, Viktor Hamburger, and Johannes Holtfreter, geneticists such as C.H. Waddington, Richard Goldschmidt, and I. Michael Lerner, and evolutionary biologists such as Ernst Mayr, George Gaylord Simpson, Sewall Wright, and Ivan Schmalhausen sought a more holistic, naturalistic approach to living processes, an approach that retained the basic scientific commitment to materialistic explanations without the limitations imposed by a literal adherence to the mechanical and reductionistic view. This second approach is sometimes characterized by the name of holism, or holistic materialism, but in its most explicit form came to be known by the end of the last century as dialectical materialism. The main difference, as I will develop in this paper, lies in the actual invoking of a dialectical process in the latter case, and its explicit lack in the former. In all other respects holistic and dialectic materialism share similar concerns and emphases, and since the later nineteenth century both have stood clearly in opposition to mechanistic materialism.

In a previous paper, I argued that strains of both mechanistic and dialectical materialism were components of Darwin's basic view of about the evolutionary process (Allen 1983). On the one hand his emphasis on organisms as individuals competing against one another in the *laissez-faire* arena of the natural environment, and his general view that the greatest good emerges from the random interaction of individual organisms, is a clear expression of mid-19th century mechanistic thinking. On the other hand, Darwin's strong sense of the interactions of parts in a complex system, and of a dialectic producing dynamic change (for example in the theory of natural selection itself, where change is brought about by the interaction of two opposing forces: heredity, or faithful replication, and variation, or unfaithful replication), express his holistic or even (implicit, at least) dialectical side.

In that same paper, I also argued that it was the mechanistic side that most predominated in the post-Darwinian period, particularly among, though not limited by any means to, neo-Darwinians. In particular, I used R. A. Fisher (1890–1962) as an example of the more mechanistically minded neo-Darwinians in the 1930s. His models of population genetics represent a clear expression of mechanistic analysis in evolutionary thought. Although the mechanistic side of Darwinian theory seemed to gain the upper hand throughout the early decades of the twentieth century, the other, more holistic and dialectical side was always present, and reasserted itself periodically, particularly in the persistent issue of the relationship between embryonic development and evolution (a topic that Fisher by-passed altogether). There has been, if you will, a dialectic of its own between mechanistic and holistic, or dialectical thinking that may well be one of the major sources of new ideas and progressive development in evolutionary theory in the 20th century.

In the present paper I would like to focus on one example of a conscious attempt to fashion a more holistic, and even dialectical, evolutionary theory in the post-Fisherian era: The work of Ivan I. Schmalhausen (1884–1963), especially his major book of 1949, *Factors of Evolution* (Schmalhausen 1949). Schmalhausen's work was highly influential not only in the Soviet Union but also in the West, though in a more limited way (Adams 1980; Wake 1986). Wake, a contemporary evolutionary morphologist, considers *Factors* to be perhaps the most comprehensive of all the great synthetic works of the 1930s and 1940s (Wake 1986:viii). Dobzhansky considered it to be "one of the 'basic books' establishing the biological theory of evolution" (quoted in Waddington 1975:98). Mark Adams, a leading historian of Russian genetics and evolutionary theory, considers Schmalhausen to be "one of the major evolutionary theorists of this century" (Adams 1980:197). And, we should not forget it was Dobzhansky who saw the English translation of *Factors* through the press and wrote the foreword to that edition in 1949.

One of the major reasons for focusing on Schmalhausen and the philosophical issues that his work addresses lies in my conviction that the examination of such questions, in a specific context, helps to sharpen our thinking about current issues in evolutionary theory. Schmalhausen addressed concerns that went beyond the Fisherian reduction of natural populations to atomistic collections of genes. He was attempting to place evolutionary questions in the context of embryonic development and genetic transmission from which they had been wrenched by the population geneticists of the 1930s. He grappled with ideas about how gene systems evolve to bring about adaption of organisms and populations as a whole; and he dealt directly with philosophical questions of reductionistic and holistic thinking that remain with us today. Schmalhausen's struggles and attempted solutions may, among other things, enlighten our present thinking about evolutionary processes.

Before proceeding, it will be useful to define several of the philosophical concepts that will recur throughout this discussion. Because of space limitations the definitions included here can be neither exhaustive nor definitive. They are meant only to provide a few basic working definitions that can be used in discussing the work of Schmalhausen and comparing it to that of Fisher.

MATERIALISM

The world view of philosophical materialism can be described as the following four general propositions:

1. Material reality exists outside of human perception and has existed prior to our knowledge of it.
2. Ideas about the world are derived from our interaction with material reality, and not from some a priori source. This is not to say, of course, that once we have developed ideas we do not impress them on reality and try to change aspects of the real world. Nonetheless, the main direction of such activity passes always from material reality to human conceptualization.
3. All change in the universe is a result of matter in motion—that is, the action of one material entity on another. The classic example of matter-in-motion is the atomic theory as elucidated in the 19th century.
4. Non-physical forces or mystical causes are inadmissable as explanations of any phenomena. For example, Hans Driesch's "entelechy" and *élan vital* (vital force) or Henri Bergson's "creative forces" in evolution would be excluded as unknowable and therefore unallowable explanations for biological processes.

Historically, there have been two major schools of materialist thought over the past 150 years: Mechanistic and dialectical materialism. It will be useful to define these two views and distinguish between them.

MECHANISTIC MATERIALISM

Mechanistic materialism is one particular form of materialism which can be summarized as follows:

1. The parts of a complex whole are distinct and separate from one another: For example, the atoms in a molecule or the gears and levers in a clock.
2. It then follows that the proper method for studying the whole is to break it down into its component parts, each of which can be investigated independently of its more complex involvement with other parts. This method of investigation is often referred to as analysis, that is, the breaking apart of the whole into its component parts.
3. Behind the method of analysis lies the general assumption that the whole is equal to the sum of its parts and no more. There are no mystical or "emergent" properties coming from the association of parts. Thus, if we know all about each part it should be possible to reconstruct the whole in its totality; nothing more is needed.
4. Systems change over time due largely to constant forces impressed on them from the outside. For example, the planets move in definable orbits because of the gravitational interaction with other bodies; organisms die through accident or through the accumulation of waste metabolites or chance mutations; populations evolve because organisms are constantly presented with challenges from a changing environment. It is important to point out that in this view change is seen as ever-present, but it does not arise necessarily out of conditions existing within the system (organism, population) itself, but through changes presented to the system by its external environment.
5. Finally, the mechanistic world view is basically atomistic, viewing phenomena in terms of a mosiac of separate, interacting, but ultimately independent parts.

DIALECTICAL MATERIALISM

Dialectical materialism shares all the basic premises of materialism, but differs in several significant ways from mechanistic materialism. Its basic propositions are contained in the five points listed below.

1. The parts of a complex whole are interconnected, and cannot be studied only in isolation from each other. One of the major characteristics of any part is its interactions within the whole. It is, therefore, not enough to study the part by itself, but the part must be studied in its dynamic interaction with other parts that make up the whole. Thus, in addition to methods of analysis, it is also necessary to devise new methods and techniques for studying component parts in their interactions.
2. From proposition 1 it follows that the whole is more than the mere sum of its component parts—it is composed of the parts plus their interactions. The emergent properties which result from the interaction of parts is nothing mystical or abstract. It is simply that the whole of a process, for example, a functioning organism, is more than the additive values of

its separate organs and tissues. There is, in a word, a hierarchical organization of matter that means that different levels show different properties.

3. Processes in the world are dynamic and developmental. Change is a fundamental part of any system, built into the interaction of the parts within the whole, and not merely impressed on the system from the outside. The simplest forms of change are wear and tear, and deterioration. But most systems change in other ways as well: for example, the development of an embryo from egg to adult, the evolution of a population from one species into two or more species, or the succession of communities on a sand dune. Dialectical materialism sees change as something more than merely the response of a system to its external environment, recognizing that systems contain within themselves the basis for their own changing states.
4. The internal forces of change within any system can be understood as a dialectic, an interaction of opposing forces, tendencies or processes. I consider it largely a moot point whether such opposing forces are actually present within the system, or simply remain as one of our own inventions for understanding and describing certain kinds of dynamic change—through debate on this very question has spilled much ink from Engels to Lenin and Althusser. As examples of the dialectical approach we can describe key metabolic processes such as growth or senescence as an interaction between anabolism (chemical synthesis) and catabolism (chemical breakdown); we can describe evolution as an interaction between the opposing forces of heredity and variation, or specialization and generality; we can describe enzymatic regulation as an interaction between constrained and relaxed allosteric states of protein molecules; and finally, we can describe the process of ecological succession as an interaction between stable and unstable associations of communities and their environments.
5. Quantitative changes lead to qualitative changes. This means that as small, or quantitative, changes within a system accumulate, eventually large-scale, or qualitative, changes or states emerge. A classic example of this principle is the boiling of water. As water is heated the temperature gradually rises (quantitative changes), but eventually somewhere between 99.999° and 100°C boiling begins (a qualitative change). A qualitative change has taken place, since steam is a different state of matter from hot, or even rapidly evaporating, water. A more biological example comes from evolutionary theory: the gradual accumulation of small variations such as base-pair substitutions, or mutations (that is, quantitative change) eventually leads to reproductive isolation and thus the formation of separate species (qualitative change). According to this principle, quantitative changes always lead, eventually, to qualitative change, thus making the evolution of new states a necessity with time. Conversely, all qualitative changes are seen to result from antecedent quantitative

changes, a relationship that only can be uncovered, in each instance, by study of the history of the process. History, then, becomes an essential component in uncovering, and thus understanding, the dynamics of any system.

The dialectical approach to processes provides a way of understanding the dynamic nature of change. It explains why change is to be expected, and why change is not largely accidental, random, or external to systems. Dialectics helps to understand how the materialist view of matter in motion leads to nonrandom, developmental (but distinctly not teleological) change. Despite prevailing views to the contrary, the method of dialects can ultimately lead to a highly rigorous understanding of how systems develop, and how one change leads to the next in a dynamic evolving process. It is not, if applied correctly, a fuzzy, general method of thinking that, as is so often claimed, simultaneously explains everything and thus explains nothing.

THE MECHANISTIC SIDE OF NEO-DARWINIAN THEORY

In his extensively documented monograph, *The Eclipse of Darwinism,* Peter Bowler discusses a number of alternative theories to Darwinism—including orthogenesis, neo-Lamarckism, and mutationism—in vogue around the turn of the century (Bowler 1983). What is apparent from Bowler's survey is that many of the objections to standard Darwinian theory pointed to the speculative or hypothetical nature of the theory of natural selection as Darwin had formulated it: for example, the assumption of usefulness of incipient stages of an organ, the chance nature of variation, the seeming impossibility of gradual change producing a qualitative change in structure (adaptation), or the apparent directionality of many evolutionary trends. The alternative theories themselves were often highly mechanistic in their formulation, for example, E. D. Cope's kinetogenesis (modification of parts through mechanical pressure), Alphaeus Hyatt's neo-Lamarckism (direct influence of the environment), Theodor Eimer's orthogenesis, or evolutionary "momentum" (a blatantly physicalist theory) and Hugo de Vries' mutation theory (where new species originate in one large discontinuous step, from parent to offspring). These theories emphasized single factors in evolution that often acted in a machine-like manner. In following this line of argument the various non-Darwinians around the turn of the century were utilizing what they took to be the modern, scientific component of Darwin's method: mechanistic, even directly mechanical, view of nature.

Nowhere is the mechanistic side of evolutionary theory better illustrated, for its subtlety and power, as well as its limitation, than in the work of R. A. Fisher. Since I have discussed Fisher in some detail elsewhere (Allen 1983), for the purposes of comparison to the work of Schmalhausen, I will only touch on the general conclusions here. As is now well established, Fisher's work in statistics and population genetics provided a major point of unification between Darwinian theory and Mendelian genetics. By treating populations largely as collections of genes whose frequencies and change in frequency over time could be studied under different conditions, computed, and even predicted, Fisher was able to provide an exact and quantitative method of analysis of the effects of selection (natural and sexual), population size, mutation, and migration, on the rate and direction of evolutionary change. His application of Mendelian concepts to the theory of natural selection provided both a quantitative and predictive means of understanding the mechanism of change in natural populations that had been lacking in the older naturalist tradition of which Darwin had been a part. Most introductory textbook accounts of the evolutionary synthesis of the 1930s present a picture of evolution that is largely based on Fisher's many articles and books, particularly *The Genetical Theory of Natural Selection* of 1930.

Fisher's approach to the fusion of Mendelian genetics with Darwinian natural selection provides a good illustration for the way in which the mechanistic materialist trend of late 19th century evolutionary theory penetrated into the modern synthesis. More perhaps than either of his contemporaries Sewall Wright or J. B. S. Haldane, Fisher was overtly and unabashedly mechanistic in his methodology. Fisher openly attempted to make evolutionary theory quantitative, rigorous, mathematical, and predictive, analagous, in his view, to the methods employed in the physical sciences. As he himself put it as early as 1922

> The study of natural selection may be compared to the analytical treatment of the Theory of Gases, in which it is possible to make the most varied assumptions as to the accidental circumstances, and even the essential nature of the individual molecules, and yet to develop the general laws as to the behavior of gases, leaving but a few fundamental constraints to be determined by experiment.
>
> **Fisher 1922:321-322.**

As historian William Provine has put it, "Fisher believed that in large populations the deterministic results of selection acting upon single gene effects reigned supreme" (Provine 1971:149). It was, indeed, Fisher's mechanistic predilection which led him, like physicists dealing with the gas laws, to emphasize the importance of infinitely large populations as the significant arenas for evolutionary change. The issue of effective population size profoundly separated the models of natural selection proposed by Fisher on the one hand and by Sewall Wright on the other. Wright's agricultural background led him to emphasize the importance, specifically, of large populations subdivided hierarchically into smaller ones with evolutionary processes occurring at multiple levels. In the small populations Wright emphasized that particular traits could become fixed by either rigorous inbreeding and selection, or by random drift.

While his mechanistic and simplifying approach provided Fisher with an extremely powerful method for dealing with evolution in a quantitative and rigorous way, it did reinforce the view that (1) genes were identical to adult traits, and therefore further divorced the study of heredity from that of embryogenesis; and (2) organisms were mosaics of independently derived traits. It was, in effect, highly atomistic and static: organisms were just bags of paired alleles, reshuffled every generation through random mating. The base-line was an infinitely large population maintained in static equilibrium by random processes—random variation and random mating. Evolution was the disturbance of this situation by such non-random conditions as assortative mating and natural selection. For example, processes of embryonic development, for example, as factors in the evolutionary process, simply did not enter Fisher's scheme.

Fisher's approach to population genetics carried atomization another step beyond the treatment of the individual as a mosaic of traits: it also treated the population as a mosaic of genes. Whereas Darwin had viewed natural populations as a group of randomly interacting individuals, Fisher viewed such populations as a group of randomly interacting genes. In Fisher's scheme the individual organism largely disappears—organisms are reduced to genetic groupings. This model has all the earmarks of the kinetic theory of gases to which Fisher's biology aspired.

To give Fisher his credit, however, he should not be portrayed as a simple-minded mechanist who made exaggerated claims for the completeness of his analytical methods. He did, after all, appreciate the enormously complex interaction of factors in natural populations, such as the role of migration, the effects of nonrandom mating, gene interactions and the like. His simplifying assumptions were absolutely essential, during the 1930s, to forge a link between Mendelian theory and Darwinian natural selection at the only level where it was possible with both the genetic and mathematical tools available at the time. Without the analytical approach, Fisher's task would have been well-nigh impossible. It is, perhaps, due to the simplified approach that Fisher took that his views gained more precedence in the 1930s and '40s than those of Sewall Wright, who dealt with more complex interactive phenomena from the start. Because of its greater complexity, for example, his shifting balance theory of evolution was always subject to considerable misunderstanding and confusion (Provine 1986:280ff). The important issue for the historian of science here is not so much exactly why Fisher took the position he did, but the fact that it was this highly mechanistic view that won the most support from neo-Darwinians and geneticists alike in the second and third decades of the 20th century, and thus brought together together two important fields—evolution and genetics—in the first phase of the "evolutionary synthesis."

TOWARD A MORE HOLISTIC APPROACH TO EVOLUTIONARY THEORY, 1930–1950: I. I. SCHMALHAUSEN (1884–1963)

While Fisher's approach, with its mechanistic and atomistic emphasis, exerted enormous influence in the generation following the publication of the *Genetical Theory of Natural Selection,* a few workers were already aware that a more holistic and dynamic approach to evolutionary change was eventually going to be necessary. These workers added to current evolutionary thought a deeper concern for genetic interactions within individual organisms, between individual organisms in a population, and between organisms and their environments, attempting in the process to construct a more thorough integration between population genetics and natural history, sytematics, paleontology and embryology than the Fisherian method provided. It is beyond the scope of this paper to discuss in any significant detail all of these latter developments. However, it will be useful, in suggesting that the mechanistic approach to evolutionary problems had its more holistic counterpart by the 1940s, to look briefly at the work of one major—and often ignored—figure, the Russian evolutionist and morphologist, Ivan I. Schmalhausen (1884–1963).

Ivan Schmalhausen's *Factors of Evolution,* first published in 1947, was reissued in 1986 in a paperback version with its original preface by the late Theodosius Dobzhansky and a current Foreword by David Wake (Wake 1986). Although introduced to Westerners by Dobzhansky in 1949, Schmalhausen's work has never gained wide recognition. In the 1940s and early '50s both Dobzhansky himself and others such as C. H. Waddington were concerned with many of the same issues as Schmalhausen: For example, the existence of cryptic variation within populations, or the evolution of morphogenesis. But, as Wake tells us, Dobzhansky was "busy making his own synthesis," and Waddington may have felt Schmalhausen's ideas too competitive with his own to advertise them broadly. Whatever the cause for the initial neglect, in the past 5 or 6 years several writers have expressed renewed interest in Schmalhausen's work (Wake 1986; Larson 1989). This interest has focused both on Schmalhausen's evolutionary concepts—in particular his ideas of dynamic and stabilizing selection—and on his openly dialectical methodology. What is this methodology and how does it differ from the more classical work of Fisher and his school?

In a review of Wake's Foreword, Mark Adams has pointed out that Dobzhansky, among others, felt that Schmalhausen's use of dialectics was primarily cosmetic, adopted to ward off attacks by the Lysenkoists, whose ascendancy to power in 1948 was ultimately to force Schmalhausen out of his major post less than two years after publication of *Factors of Evolution* (Adams 1988). I disagree with that interpretation. A close reading of Schmalhausen suggests that he was genuinely thinking in dialectical terms, consciously trying to apply the dialectical method to a range of problems that had been largely ignored by Western geneticists and evolutionists during the early years of the synthesis. As Loren Graham demonstrates so clearly, a broadly-based group of Soviet natural and social scientists were actively developing dialectical materialism as an open and explicit philosophy of science, applied to their own research, from the early 1920s onward (Graham 1986). In particular, in the field of Darwinian theory, dialectics found a hospitable and fertile ground for expansion. Ironically, T.D. Lysenko (1898–1976) denounced Schmalhausen in 1948, even though, as Levins and Lewontin have noted, Lysenko's work, rather than Schmalhausen's, represented the misapplication of dialectics. Schmalhausen, for example, was the one who saw the correct dialectical synthesis between heredity and variation but he was denounced by those who got it wrong (Levins and Lewontin 1985).

Schmalhausen's commitment to a dialectical viewpoint is evident first and foremost in his interest in developing a true synthesis between genetics and evolutionary theory, on the one hand, and genetics, evolution, and embryonic development, on the other. Schmalhausen related evolution to a variety of embryological/developmental problems attempted by only a few other investigators at the time: The problems of nucleo-cytoplasmic interaction (he cited the work on *Acetabularia,* by J. Hämmerling), on embryonic induction and organizers (the Spemann School), and the concept of differentiation through the regional effects of metabolic gradients, (the work of C.M. Child).

In a number of ways Schmalhausen continually emphasized the dialectical process in terms of the contradictory forces involved in evolutionary change: the opposing forces of heredity and variation, organism and environment, individual and population, flexibility and specialization, internal and external, etc. In this respect, as well as in some others, Schmalhausen was similar to his contemporary C. H. Waddington (1905–1975), who also explicitly claimed to use dialectics as a conscious method (Waddington 1975). Both Waddington and Schmalhausen believed that dialectics was a highly useful methodology, providing an analysis that tied together many facets of evolutionary biology. That the method can be useful and is not necessarily invoked as political window dressing is supported by the fact that Waddington and other Western scientists, such as J. B. S. Haldane or J. D. Bernal, tried to put dialectics into practice when there was no political necessity for—and certain professional risks involved in—doing so. There is thus good reason for maintaining that Schmalhausen's use of dialectical materialism was a genuine attempt to apply a new methodology and approach to evolutionary questions, and not merely a political move as Dobzhansky seemed to believe.

To understand Schmalhausen's use of a dialectical method, we will consider his concept of norm of reaction and the two opposing forces of dynamic and stabilizing selection. A norm of reaction represents the basic level of variability, or range of expression, of which every genotype is capable. As Schmalhausen put it, "every genotype is characterized by its own specific 'norm of reaction,' which includes adaptive modification of the organism to different environments" (Schmalhausen 1949:7–8). Norms of reaction are the inherited ranges of response of which every genotype is capable, and are the product of natural selection acting on the whole genotype (individual loci along with their various epistatic interactions). A norm of reaction can be thought of as represented by a normal (bell-shaped) curve of distribution around a mean. The various points on the curve represent the expression of one given genotype in different environments (not different genotypes in the same environment) and thus demonstrate the built-in flexibility characteristic of a particular genic system. In Schmalhausen's formulation, the norm of reaction has been selected for as a means of allowing the organism (and the species) to respond adaptively to various environmental conditions. Thus, for example, a norm of reaction in mammals allows the various muscles to increase in size with use, or decrease in size with non-use. This ability has a range of possibilities beyond which, genotypically, it does not go: Muscles are not infinitely expandable by persistent exercise, nor do they totally degenerate with little use. This norm of reaction or range of possible developmental responses is built into the genotype for mammalian muscle structure and is the product of a long history of selection.

The norm of reaction for any trait may be characterized by a large or small amount of variance—that is, the bell curve can be either wide or narrow. Each shape would represent a different potential for adaptive responses to the environment in which the species lives: broad norms of reaction would be advantageous in more variable environments, while narrow norms of reaction would be advantageous in less variable environments. In Schmalhausen's terms, norm of reaction is directly related to adaptive responses which are historically the product of selection within a particular environment. The norm of reaction for any trait is not a genetic process independent of environment, or an accidental genetic combination resulting from a recent event such as hybridization or mutation. It is very important to understand this concept in terms of its historical nature and of its dynamic, interactive nature. The norm of reaction in any particular instance is the product of a constant interaction between phenotype and genotype, on the one hand, and the organism and its environment, on the other. But how is a norm of reaction established? To explain this process Schmalhausen invokes the concepts of dynamic and stabilizing selection—a pair of dialectical opposites that work continually to both change and retain the existing norm of reaction for any characteristic. We turn now to a brief examination of these two forms of selection.

Dynamic selection is the more conventional selection process. The most usual type of dynamic selection occurs among organisims that live in a constantly changing, nonhomogenous environment. An example of such an environment is one with seasonal variations in temperature or moisture and characterized genetically by polymorphism—the kind of model Dobzhansky was already investigating in the American southwest with *Drosophila* chromosome variations as early as 1946 (Dobzhansky 1947). Schmalhausen recognized two kinds of genetic traits: labile and stable. Labile traits are those that vary in expression in different environments, while stable traits are those that show little variation even under large differences in environment. Among labile traits, Schmalhausen recognized two types: Morphosis and modifications. Morphoses represent changes beyond the norm for the population, and are neither inherited directly in future generations, nor are they adaptive; these are mere accidents or erroneous variations in embryogenesis resulting from some random environmental change (temperature or moisture). Phenocopies are examples of morphosis (Schmalhausen 1949:175). Modifications are changes beyond the norm that may or may not be inherited directly, but that are definitely adaptive. An example of modifications is found in genetically different strains of flax which, when grown in new environments, phenotypically come to resemble flax native to those environments.

Stable traits, on the other hand, are more rigidly controlled by internal factors—that is, genes—and thus are less dependent for their expression on environmental factors. The importance of making the distinction between labile and stable traits for Schmalhausen was to emphasize that, contrary to an underlying assumption of the Fisherian school, in most cases there was no one-to-one correspondence between genotype and phenotype (Schmalhausen 1949:76). In other words, the interaction of genes within the genotype, or the entire genotype with the environment, produces a new hierarchical level of development in which new properties emerge. The whole notion of interactions producing emergent properties—an essential feature of dialectical materialism—is important in that it allows Schmalhausen to get at a much more interesting point, namely, the role of stabilizing selection in evolution.

Stabilizing selection is that aspect of natural selection that brings about the conversion of labile traits (as long as they are adaptive) into stable ones. This must occur through the action of selection on morphogenetic processes, continually internalizing external cues and thus regularizing their effects on the developmental process. According to Schmalhausen, the stability of the morphogenetic system is rendered labile by variation in the environment or the germ plasm, or both. Thus, in a highly variable environment there is constant change in the norm of reaction within a population (Fig. 1). This change is the result of selection acting on small, inherited variations. However, such changes produce a disruption in the integrated morphogenetic process. Stabilizing selection is continually at work, selecting for any changes in morphogenesis that lead to a dependence on internal, or regular rather than external, or irregular cues. All organisms must be able to adapt to changing conditions, so that some reliance on external cues is vital. However, large-scale adaptive

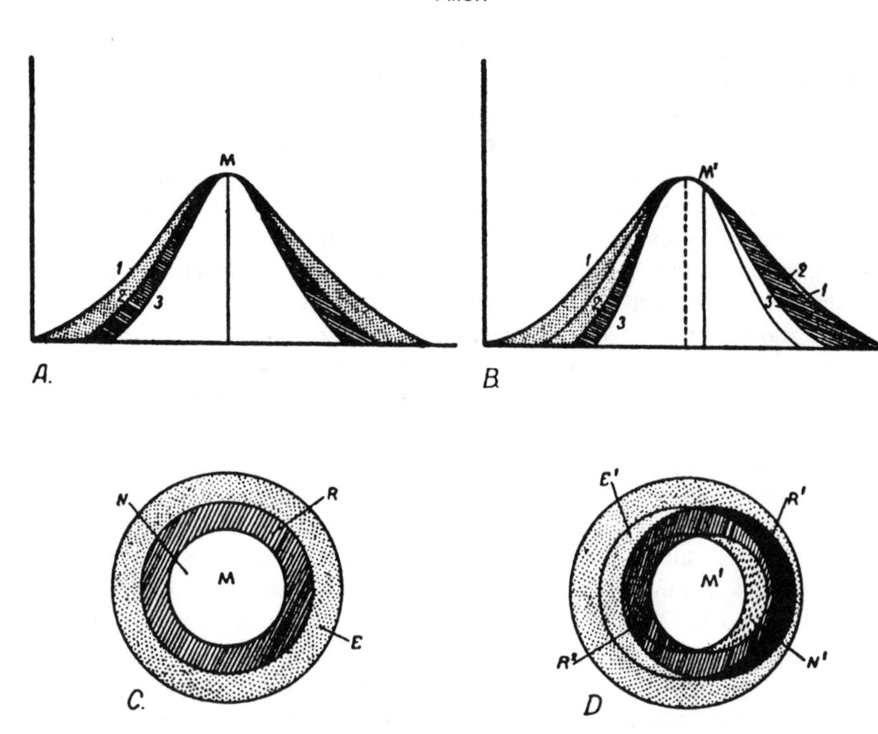

Fig. 1. Selection in a heterogeneous and variable external environment (taking into account different forms of individual reaction, *i.e.*, modifications). (A) Selection under fixed (in a given heterogeneity) conditions (stabilizing selection). Frequency curve: (1) Of all individuals; (2) of individuals that survive under actually encountered diverse conditions (conditionally normal individuals, conditionally favorable and partially favorable variations; (3) of individuals surviving and leaving offspring under the usual average "normal" situation. The stippling denotes the zone of absolute elimination of deviation that is harmful under all conditions (mutation and morphoses). The oblique hatching denotes the zone of variations, which are under common normal conditions, but which survive and leave offspring under definite, variable but actually encountered conditions. (B) Selection when these conditions vary in a definite direction (with persistent heterogeneity). Dynamic selection. Frequency curve: (1) Of all individuals; (2) individuals surviving under the different conditions of the new environment; (3) individuals surviving under the new "normal" situation. (C–D) Graphic illustrations of the mechanism of stabilizing and dynamic selection (as in A and B).

changes are favored if they can be internalized—that is, stabilized—and no longer depend on more haphazard external signals.

To explain how such changes could have evolved, Schmalhausen used as an example the general adaptation of plants to drought conditions. Most plants, he pointed out, develop leaves that function during periods of sufficient moisture, but then are shed during periods of drought or cold. Xerophytes (desert plants), such as members of the genus *Acacia,* have evolved an adult form that lacks true leaves altogether. The plant has leaf-like organs, called phyllodes, that develop not from whole leaf primordia but from the petiole. In xerophytes the phyllodes, along with the stem, have a much greater number of stomata and chloroplasts than normal petioles and stems, and carry out most of the plant's photosynthesis. During morphogenesis, Schmalhausen shows, *Acacia pycnantha* at first develops pinnately compound leaves typical of the genus as a whole, but these are soon replaced by the modified petiole, or phyllodes (Fig. 2). Evolving in a desert environment, ancestors of *Acacia* may well have initially produced pinnate leaves that survived only the spring rains, and then were shed in response to increasing aridity. Development in these ancestral forms was largely governed by an environmental cue. Variations in the morphogenetic process that would link leaf transformation to an internal cue would have an advantage: the plant would not have to go through a period

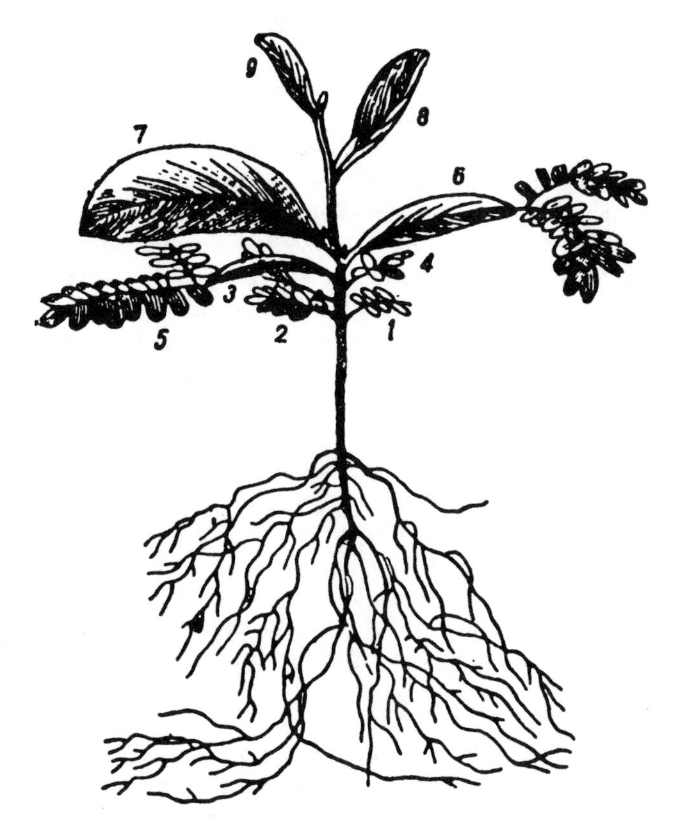

Fig. 2. Young plant of Acacia pycnantha. The initial pinnately compound leaves are replaced by transitional leaves (5–6) which develop later into typical phyllodes (7–9). (Courtesy, K. Zimmermann, "Die Phylogenie der Pflanzen," Fig. 248, Jena, 1930.)

of physiological response to desiccation, but could undergo a developmental change that anticipated the environmental change. Dynamic selection favors any genetic variations that would reduce the plant's exposure to desiccation and water loss: for example, responding to an earlier environmental cue such as temperature increase, or day-length. However, stabilizing selection would favor any genetic variations that could convert the plant's general ability to respond to dryness into a specific characteristic in which the leaves are replaced by phyllodes early in morphogenesis. Stabilizing selection would thus act to internalize and fix the trait by incorporating it into the developmental process. Dynamic selection, on the other hand, would work to favor variations in morphogenesis such as broader phyllodes or increased numbers of chloroplasts for increased photosynthesis, etc.

The distinction between dynamic and stabilizing selection can be understood most fully by viewing their respective effects on the norm of reaction. In the Acacia example, dynamic selection would act on the existing genotype for leaf shape, favoring any variation, or morphosis, that would reduce leaf surface area by altering the process of leaf development. Stabilizing selection would also be acting on the genotype, simultaneously, to stabilize phyllodic leaf shape—that is, to change the norm of reaction so that phyllodic shape becomes more and more fully penetrant (in the genetic sense) so that it becomes more fully expressed even when environmental cues are absent (that is, under a broader array of environmental conditions). It is important to see that in Schmalhausen's scheme both dynamic and stabilizing selection are essential to the evolution of any given norm of reaction. If only dynamic selection were involved, the desired new state, or norm of reaction, might be quite variable in its penetrance or expressivity against different environmental and genetic backgrounds. In other words, any new genetic variations, even if adaptive, might vary widely in how the genotype was expressed in a given range of environments. Stabilizing selection acts on the genotype to increasingly internalize the cues on which the developmental process depends (that is, the processes by which genotype is connected into phenotype). Stabilizing selection thus ensures that the beneficial phenotype is more fully expressed in spite of potential variations in the environment in which the organism develops. Dynamic and stabilizing selection go hand-in-hand to the extent that a positive trait can be established by dynamic selection most effectively if its expression is constantly being internalized by stabilizing selection.

Another illustration of the interaction between stabilizing and dynamic selection is the evolution of the powerful masticating and the temporal muscles, in conjunction with progressive development of the dental system, jaws, cheekbone, and sagittal ridge in predatory mammals. Here Schmalhausen pointed out how dynamic and stabilizing selection could interact in the transformation of an insectivorous to a carnivorous lifestyle. The first result of this transition, he argued, was probably "only an intensified exercising of the maxillary muscles so their mass, their blood supply, and the surface of their attachment to the skeletal parts were enlarged," (Schmalhausen 1949; 1987). This came about simply by increased exercise of the jaw muscles due to chewing meat and tough animal tissues. These changes were adaptive modifications such as come from exercising any muscle. As such they would not be inherited. However, Schmalhausen shows that change in muscle size through exercise comes about by enlarging individual fibers, not increasing the number of fibers. Any genetic variation in the developmental process that would favor an actual increase in number of muscle fibers moving the jaw would have a selective advantage in that it would increase the animal's power of chewing. Such increased number of cells would be the result of dynamic selection. Thus, what was initially an external cue—stimulation of muscle growth through additional use—eventually becomes internalized as larger muscle growth during ontogeny. Once an increase in the number of muscle fibers occurs, it can be acted on by stabilizing selection, conferring on the offspring a selective advantage in the struggle for existence. Stabilizing selection would come into play by fixing those genes that control the number of muscle fibers—that is, by making a permanent part of ontogeny the genotype for an increased number of muscle cells operating the jaws. Similarly, dynamic and stabilizing selection would also act on other genes that regulate the size and shape of the various skull bones, including the sagittal crest where jaw muscles are attached. Dynamic and stabilizing selection act on multiple gene systems, since any norm of reaction—indeed any trait as Schmalhausen sees it—is the result of the interaction of a number of genetic loci.

The emphasis that Schmalhausen gives to interactions between genes, like his emphasis on the interaction of dynamic and stabilizing selection, seems to me a clear illustration of his commitment to dialectical materialist thinking.

In Schmalhausen's terms, evolution is the result of a constant dialectic between the action of dynamic and stabilizing selection. Dynamic selection leads to morphogenetic changes that are adaptive to new environments; stabilizing selection internalizes and adjusts their norms of reaction, tying them to internal cues. The interaction of such selective processes produces new, stable states of morphogenesis that are associated with major new directions in evolution. A series of quantitative changes (e.g., selection for the establishment of developmental cues in embryogenesis) has led to a qualitative change (development of an internal and predictable morphogenetic change due to internal cues).

SCHMALHAUSEN'S DIALECTICS

A full reading of *Factors of Evolution* makes it abundantly clear that dialectical materialist thinking lies at the core of what Schmalhausen saw to be the correct approach to evolutionary thinking. First, Schmalhausen clearly formulates the dynamics of evolution in terms of opposing forces or tendencies: Dynamic vs. stabilizing selection, heredity vs. variation, periods of genetic change vs. periods of stasis, internal cues vs. external cues in the regulation of morphogenesis, etc. To modern readers this may appear to be just logical or empirical thinking rather than dialectics per se, but the unity and consistency of Schmalhausen's thought argues for the existence of a more conscious and well-thought-out philosophical approach. Most dialectical materialists emphasize that any good, clear observer and thinker will ultimately come to a kind of dialectical thought, even without hearing of the work of Marx, Engels or Lenin. Schmalhausen does indeed make updated references to dialectics and to the work of Engels, among others, in describing aspects of the evolutionary process (see, for example, 1949:xix; 61). But it is the full integration of dialectical thinking—particularly his view of genetics as a synthesis of heredity and variation—in his description of the evolutionary process that undermines the view that he was only engaged in political rhetoric.

Wake and Larson have both pointed to the important role that hierarchical thinking plays as an aspect of dialectics in Schmalhausen's philosophy (Wake 1986; Larson 1989). Levels of organization form a crucial aspect of the dialectical process of quantitative change giving rise to qualitative change. As cells associate to form a higher level of organization, a tissue, for example, new properties emerge that are not apparent in the cells themselves. There is nothing mysterious about the emergent properties, though the issue of emergence was very much debated in the 1920's and 1930's, especially in reference to the vitalism of C. Lloyd Morgan (1852–1936), and the concept of emergent evolution of Henri Bergson (1859–1941). At the same time, in any hierarchy, there is nothing predictable about the properties of higher levels from knowledge only of lower levels. Schmalhausen saw the developmental process as embodying hierarchical organization, since embryogenesis, for example, involves the formation of hierarchies of organization as undifferentiated cells give rise first to tissues then to organs and finally to organ systems. This process of hierarchical development is itself, in Schmalhausen's view, a product of the evolutionary process. Furthermore, as organisms evolve complex ecological relations with their environment and other species, a still higher level of organization—the populational or community—emerges. Lower levels of organization, for example, the genotype of a population, determine higher levels—that is, general adaptive phenotypes—but not simply in an additive way (Schmalhausen 1949:xxi). Permeating Schmalhausen's work is his recognition of the importance of understanding levels of organization in any system and the role of emergent properties.

CONCLUSIONS

In this paper I have tried to show the difference between the more mechanistic approach to evolutionary theory as represented by the population genetics of R. A. Fisher, and the more dialectical and holistic approach as exemplified by the concepts of norm of reaction and dynamic and stabilizing selection put forward by Ivan I. Schmalhausen. Although nearly contemporaries, the two men represent very different perspectives and world views regarding the evolutionary process. Fisher, heir to the British empirical and mechanical tradition stemming back as far as Bacon, Boyle, and Newton, emphasized atomism, the discreteness of parts, and the additive nature of interacting processes. Schmalhausen, heir not only to the more holistic tradition of 19th century German and Russian philosophy, but also to the newly developed tradition of dialectical materialism in the Soviet Union in the twentieth century, sought to apply those traditions to the development of a more holistic and interactive evolutionary theory.

There are several conclusions I would like to draw from this study of Schmalhausen and his place in the evolutionary synthesis. The first is that there has been in the past a dialectic evident in the history of evolutionary theory itself: That is, an opposition between the mechanistic and analytical approach to evolutionary theory on the one hand, and the dialectical or holistic approach on the other. Far from being undesirable, both of these approaches have their place in the ongoing dynamic of evolutionary study. The mechanical or analytical view can be useful at the initial stages of study of any complex phenomena. It is impossible to look at a complex process as a whole, and only as a whole, and expect to understand the details of its workings. At the same time, to employ only the mechanistic or analytical approach is to miss many of the interactions that make the whole of any process more than the sum of its parts. Thus, the emergent properties resulting from changes in the embryogenesis of an interrelated set of structures (jaw muscles, points of attachment, etc., for example) would be missed by looking only at changes in gene frequencies for one or at most a few genes at a time. The present paper suggests that we need to employ both mechanistic and dialectical, or holistic, ways of viewing any natural process. Ideally, each individual investigator should be able to combine both viewpoints, and look at a problem from both perspectives. At the very least, however, we should encourage the kind of dialectic between such points of view amongst the members of a research community. The real danger comes from the claim that either philosophical position alone will provide a complete understanding of any natural process.

The second conclusion is that evolutionary theory still needs to complete its synthesis by including the processes of morphogenesis (embryonic development). Although this point has been made many times over the years, relatively few concrete examples exist of how such a synthesis might be accomplished. I have tried to show that Ivan Schmalhausen offered one of the more productive attempts to bring embryonic development into the evolutionary synthesis in the 1930s and 1940s. Hopefully, increasing numbers of evolutionary biologists will become familiar with his work in some detail and carry it further than was possible in 1949.

There is much to be gained by effecting this part of the synthesis at the present time. For one thing, it is becoming increasingly important to understand in more detail the interrelationship between the genotype and the cellular and external environments in which organisms develop. Understanding the morphogenetic processes by which genotypes become phenotypes can help us avoid identifying adult phenotypes with specific genotypes. Among other issues that such study could clarify are the currently popular but ill-conceived claims that complex human behaviors such as alcoholism or criminality have a significant genetic component. By essentially ignoring the entire developmental process, which in terms of human psychology and behavior is even more dramatic than in other animals, such claims present a naive and highly mechanistic account of human behavior. A more dialectical approach to evolutionary thinking that brings ontogeny into the picture would in fact help to stem the tide of such oversimplistic claims.

The third and final conclusion I would like to make is perhaps self-serving, but important nonetheless. I think that the study of history and philosophy of science has a great deal of value for the current practice of science itself. Research strategies and approaches can be improved, I would argue, by understanding explicitly the philosophical bases of our own as well as others' arguments. Just as we can sharpen our methodologies with regard to laboratory techniques and instruments, we can also sharpen our methodologies with regard to research strategies and approaches to complex problems. In this vein I am calling for increased interaction between historians and philosophers of science, on the one hand, and practicing scientists on the other, as a process that by expressing its own dialectic would be mutually beneficial to both.

ACKNOWLEDGMENTS

I am much indebted to Allan Larson of the Biology Department at Washington University, for his many helpful suggestions on this paper. My understanding of Schmalhausen is much improved through his thoughtful criticisms.

REFERENCES

Adams MB (1988): A missing link in the evolutionary synthesis. Isis 79:281.

Adams MB (1980): Severtsov and Schmalhausen: Russian morphology and the evolutionary synthesis. In Mayr E, Provine WB (eds): "The Evolutionary Synthesis." Cambridge (MA): Harvard Univ Press, p 193.

Allen GE (1969): Hugo De Vries and the reception of the "Mutation Theory." J Hist Biol 2:55.

Allen GE (1978a): "Thomas Hunt Morgan. The Man and His Science." Princeton: Princeton Univ Press.

Allen GE (1978b). "Life Science in the Twentieth Century." New York: Cambridge Univ Press.

Allen, GE (1980): "Dialectical materialism in modern biology." Science and Nature 3:43.
Allen GE (1983): The several faces of Darwin. In Bendall DS (ed): "Evolution from Molecules to Men." Cambridge: Cambridge Univ Press, p 81.
Allen GE (1990): From embryologist to geneticist: T. H. Morgan and the influence of mechanistic materialism in the development of modern biology, 1890–1940. Amer Zool.
Anonymous (1978): Obituary: I. Michael Lerner. Genetics 88 (Suppl):139.
Barrett PH, Weinshank DJ, Gottleber TT (1981): "A Concordance to Darwin's Origin of Species, First Edition." Ithaca, NY: Cornell Univ Press.
Cope ED (1904): "Primary Factors of Organic Evolution." Chicago: Open Court Publ.
Darwin C (1859): "On the Origin of Species." London: John Murray. [Facsimile edit: Cambridge (MA): Harvard Univ Press, 1964.]
Dennert E (1903): "Vom Sterbelages Darwinismus." Stuttgart.
Dewey J (1909): The influence of Darwinism on philosophy. In Dewey J: "The Influence of Darwinism on Philosophy and Other Essays." New York: Henry Holt and Co., Inc.
Dobzhansky T (1947): Adaptive changes induced by natural selection in wild populations of Drosophila. Evolution 1:1.
Ehrlich PR, Raven PH (1964): Butterflies and plants: A study in co-evolution. Evolution 18:586.
Eimer T (1888): "Die Entstehung der Arten auf Grund Von Vererbung erworbener Eigenschaften." Jena: Gustav Fischer.
Fisher RA (1915): The evolution of sexual preference. Eugenics Rev 7:184.
Fisher RA (1922): On the dominance ratio. Proc R Soc Edinburgh 42:321.
Fisher RA (1930): "The Genetical Theory of Natural Selection." Oxford: Clarendon Press.
Ford EB (1940): Genetic research in the Lepidoptera. Ann Eugenics 10:227.
Gillespie C (1960): "The Edge of Objectivity." Princeton: Princeton Univ Press.
Gould SJ (1980): Is a new and general theory of evolution emerging? Paleobiol 6:119.
Graham L (198): "Science, Philosophy and Human Behavior in the Soviet Union," 2nd ed. New York: Columbia Univ Press.
Hogben L (1927): "Principles in Evolutionary Biology." Capetown, South Africa: Juta.
Johannsen W (1911): The genotype conception of heredity. Amer Naturalist.
Larson A (1989): The relationship between speciation and morphological evolution. In Otte D, Endler J (eds): "Speciation and its Consequences." Sunderland, MA: Sinauer.
Lerner I (1954): "Genetics Homeostasis." New York: John Wiley and Sons.
Levins R, Lewontin R (1985): "The Dialectical Biologist." Cambridge, MA: Harvard Univ Press, Chapter 7.
Lewontin R, Levins R (1978): Evoluzione. In Enciclopedia V:995.
Loeb J (1912): "The Mechanistic Conception of Life." Chicago: Univ Chicago Press. [Reprint edit, Cambridge (MA): Harvard Univ Press, 1964.]
MacKenzie DA (1981): "Statistics in Britain, 1865–1930." Edinburgh: Edinburgh Univ Press.
Mather K (1943a): Polygenic inheritance and natural selection. Biol Rev 18:32.
Mather K (1943b): Polygenics in development. Nature 151:960.
Mayr E, Provine W (1980): "The Evolutionary Synthesis." Cambridge (MA): Harvard Univ Press.
Provine W (1971): "The Origins of Theoretical Population Genetics." Chicago: Univ Chicago Press.
Russett C (1966): "The Concept of Equilibrium in American Social Thought." New Haven: Yale Univ Press.
Sachsse H (1968): "Die Erkentnnis des Lebendigen." Braunschweig: Freidrich Vieweg & Sohn.

Schmalhausen I (1949): "Factors of Evolution." Philadelphia: Blakiston. [Reprint edit, Chicago: Univ Chicago Press, 1986.]

Schweber S (1980): Darwin and the political economists: Divergence of character. J Hist Biol 13:195.

deVries H (1901–1903): "Die Mutationstheorie." Leipzig: von Veit & Co. 2 vols.

Waddington CH (1942): Canalization of development and the inheritance of acquired characters. Nature 150:563.

Waddington CH (1948): The concept of equilibrium in embryology. Folia Biotheoretica 3:127.

Waddington CH (1975): "The Evolution of an Evolutionist." Edinburgh: Edinburgh Univ Press.

Wake D (1986): "Foreword to Factors of Evolution by I. I. Schmalhausen." Chicago: Univ Chicago Press: V.

Through the Looking Glass: The Evolution of Soviet Darwinism

MARK B. ADAMS
History and Sociology of Science, University of Pennsylvania, Philadelphia, Pennsylvania 19104-6310

> The problems frequently and the mode of attack nearly always are quite different from those in other countries and where they have followed an influence from abroad (as in the Drosophila work) they have given it a direction and a method that is peculiarly local or Russian. Aside from this peculiarity, they are nearest in spirit and sympathy to the American type.
>
> **L. C. Dunn to Hutchinson, 2 November 1927**

INTRODUCTION

Much has been written about the history of Darwinism in Britain and America, but the literature on the rich evolutionary traditions of other countries has been remarkably sparse, and all too much of it has sought to explain why their work lagged so far behind our own. We do not know much about the way biologists in France, Germany, or Russia thought about the evolutionary process, and until we do, no matter how world-class our science, our perspective will remain essentially provincial.

The development of evolutionary theory in Russia is of special interest for several reasons. The first is that its traditions helped to shape our own. Russian scientists dominated experimental population genetics for more than thirty years (Chetverikov, Timoféeff-Ressovsky, Dobzhansky), and they made major contributions to studies of biochemical evolution (Vernadsky, Kol'tsov, Oparin), plant evolution (Vavilov, Levitsky, Karpechenko), and ecology (Sukachev, Gause). Indeed, in the list of the founders of the synthetic theory of evolution compiled by G. G. Simpson (1953), almost a quarter are Russian. To understand their contributions, then, is to understand some important sources of our own thinking.

There is a second, more immediate reason. Although Western evolutionary theory was invigorated by interaction with Russian thought until around 1930, the imposition of Stalinist barriers to migration and the open flow of ideas led to the isolation of Soviet scientists from their Western colleagues. Despite the subsequent zigs and zags of international and Soviet politics, Russian scientists have yet to regain the full freedom of international communication, publication, and travel so central to the way modern science works. Today there is hope that things may be changing. If barriers disappear and Soviet scientists become reintegrated into international scientific life, they will be bringing with them a unique historical legacy, a scientific past quite different from the one we have known. We will be in a better position to appreciate our new colleagues if we understand where they are coming from.

Finally, the Russian case may help us to understand how science develops. *Isolation* and *barriers to migration* are terms familiar to students of organic evolution. But organisms are not the only things that evolve: historians and philosophers have recently found evolutionary or natural selection models especially useful in studying the history of science and technology (Kuhn 1962, Toulmin 1972, Adams 1979, Richards 1987, Hull 1988, Basalla 1988). Darwin found comparative biogeography crucial to his understanding of the evolution of life; we who study the evolution of science would do well to follow his example. And as Sewall Wright has pointed out (1931), organic evolutionary change proceeds most rapidly in small to middle-sized semi-isolated populations (although much of it does not enhance fitness). When applied to the history of ideas, Wright's insight suggests that we should be sensitive to the populational structure of the scientific enterprise and to the relationship between marginality, isolation, and change.

In this context, the unusual isolation of Soviet evolutionary theorists for several decades is an anomaly of considerable interest. For, as we will see, instead of providing a more or less distorted reflection of our own scientific development, the Soviet case affords us a glance through the looking glass into a place where the history of science temporarily took a different route, where evolutionary science became an enterprise somewhat different from the one we know. As Alice found, such a world may appear strange from our everyday perspective, but understanding its special logic can help us to see our own world in a fresh way.

DARWINISM EAST AND WEST

Darwinism is not quite the same thing as biological evolutionary theory. Analytically, their semantic territories overlap, of course, but there are many theories of organic evolution that are not "Darwinian," and, conversely, the term "Darwinism" encompasses social and philosophical meanings that go well beyond science. Indeed, one historian has counted at least a dozen meanings of Darwinism, including a theory of organic evolution, a philosophy of science, an ideology of social progress, and a world view—and he was limiting himself to the English-speaking world (Greene 1981). Historically, the relationship between Darwinism and the theory of biological evolution is even more complicated: although it is commonplace to point to evolutionary theories antecedent to Darwin's (e.g., Lamarck's), recent historical work is demonstrating that the term *Darwinism* came into widespread use (in the 1870s and 1880s) some years earlier than the terms *evolution* and *biology* assumed their modern meanings. The distinction between Darwinism and evolutionary theory is an important one, because in Russia (and perhaps elsewhere) disputes over the meaning and proper relationship between the two helped to define discourse and discipline.

"Like everything else," historian John Greene has reminded us, "Darwinism evolves" (Greene 1981). Thanks to recent studies, the outlines of that evolution are becoming clearer. As Darwin's ideas migrated into different national settings, they underwent mutation, recombination, and selection, adapting to local scientific, political, and cultural conditions so as to produce distinct varieties of Darwinism—which is why the meaning and connotations of the term vary from country to country.

Late Tsarist Russia had distinguished traditions in zoology, botany, and geography that had been shaped by the country's rich natural history and the exploration of its vast territories. Even before Darwin, many Russian naturalists had adopted transformist theories, derived variously from French *transformisme* and German *Naturphilosophie.* In the 1850s, Russian losses in the Crimean War discredited the autocratic, orthodox, isolationist ideology of Nicholas I, leading to a new national interest in Western science, technology, economics, and values. The introduction of Darwin's theory into Russia in 1860 coincided with the era of the Great Reforms of Tsar Alexander II, which initiated a quickening of culture and a half-century of lively social ferment. In Russia (as elsewhere), Darwin's ideas were freely appropriated by diverse groups to lend scientific legitimacy to their varied social agendas. But if Russian intelligently were generally enthusiastic about Darwin's theory of "descent with modification" (as he called it in the *Origin*), they were skeptical about a natural selection based on the intraspecific struggle for existence occasioned by Malthusian overpopulation, a concept that seemed more suited to the industrialized island of Britain than to the underpopulated expanses of the Russian Empire. This special feature of the Russian variety of Darwinism in the nineteenth century has been described and explained in an important new book (Todes 1989).

Around the turn of the century, there was a growing conviction in many countries that Darwin's theory had outlived its scientific usefulness. The traditional primacy of natural history was being challenged by the emergence of *biology* as a new, experimental science of life, subdivided into fields not by the kind of organism being described (e.g., zoology, entomology, ornithology), but rather by the kind of universal life-process being experimentally manipulated (e.g., heredity, development, reproduction). If biology was to take its rightful place alongside the other experimental sciences of chemistry and physics, many biologists felt, it could no longer tolerate the sort of speculative philosophizing that Darwin and his followers, however great, had indulged in.

The result has been described as an "eclipse of Darwinism" (Bowler 1983), but the term can be misleading. Almost without exception, biologists had come to accept the reality of organic evolution: Darwin's theory of descent had won the day. Not so natural selection, which seemed altogether too hypothetical a mechanism to meet the new laboratory canons of rigor. Thus, even during the eclipse, most biologists treated Darwin's name with great respect: having mobilized natural history to demonstrate once and for all the reality of evolution, he had left to the new biology the problem of finding its cause.

By the late nineteenth century, these European trends had taken root in St. Petersburg, Russia's "window on the West," and up-to-date biologists there shared growing Continental reservations about natural selection. Western evolutionary skeptics were widely translated, and Henri Bergson's philosophy of *évolution créatrice* was in vogue. By the outbreak of World War I (when it was renamed Petrograd), the Russian capital had already given birth to one non-Darwinian theory of evolution (Sergei Korzhinsky's heterogenesis, a variant antecedent of De Vries's mutation theory) and over the next decade it would sire several more.

In Moscow, however, things were different. Its university had long been a stronghold of transformism and later Darwinism, spearheaded first by N. A. Severtsov and later by his student Mikhail Menzbier, whose studies of bird systematics, distribution, ecology, and behavior were pursued in a Darwinian spirit. Menzbier's perspective was passed on to his many students, among them A. N. Severtsov (the son of Menzbier's teacher) and Nikolai Kol'tsov. Severtsov's German graduate training turned him into a consummate morphologist convinced that the chief task of the evolutionist was to chronicle the evolutionary history of life on earth in order to determine its broader patterns; for him, morphology was *the* evolutionary science (Adams 1980c). By contrast, Kol'tsov's European trek a decade later converted him to experimental laboratory biology and the new "cellular" morphology practiced at the Stazione Zoologica in Naples and the Russian marine station in Villefranche. True to his Moscow heritage, however, Kol'tsov remained a confirmed Darwinian, holding that traditional training in zoology and botany was a prerequisite for experimental excellence (Adams 1980a, 1980b).

Prior to 1915, then, Russian Darwinism was a variety of European. Although it had assumed particular characteristics by virtue of Russia's natural, scientific, cultural, and sociopolitical history and setting, its distinctness was limited by the intelligentsia's ever-increasing contacts with the West. Since the 1860s Russians had traveled, studied, worked, and published abroad with relative ease. Russian journals included lengthy summaries in German or

French, and many European scientific books and articles were published in Russian translation yearly. Thus, since the late 19th century, Russians had played an active (if marginal) role in Continental scientific life.

With the outbreak of World War I, however, Russians became increasingly isolated from Western science. Although some information filtered into Petrograd, less found its way to Moscow, and Kiev was largely cut off. In February 1917 Tsar Nicholas II was overthrown, and in October the Bolsheviks seized power, leading to four years of bloody civil war during which time disease, starvation, and political executions took an appalling toll. In 1921, faced with widespread starvation and economic collapse, Lenin replaced War Communism with his New Economic Policy, restoring private ownership to many areas of the economy; things began gradually to recover.

In late 1921, then, Russian scientists emerged from seven years of isolation to learn what they had missed. Western scientific relief efforts included food, clothing, money, and vast quantities of free books and journals. For Russian life scientists, the big news from abroad was that experimental biology had been largely triumphant, and one of its most remarkable achievements was the new chromosomal theory of heredity embodied in an American book, *The Mechanism of Mendelian Heredity* (1915), by T. H. Morgan and his Columbia colleagues. Almost a decade of isolation, followed by a sudden influx of these new ideas, produced a remarkable decade of hybrid vigor.

THE GOLDEN AGE

In the early 1920s, the vigorous and diverse scientific life of late Tsarist Russia reasserted itself with new energy. True, four years of revolution, civil war, and famine had decimated the scientific establishment and led to the death or emigration of much scientific talent; on the other hand, new places opened for the younger generation, accelerating prerevolutionary developmental trends without altering their direction. The tensions between traditional natural history and experimental biology were still in evidence, but both styles prospered; Petrograd became an even more productive center of non- or anti-Darwinian theorizing; in Moscow the Darwinian tradition continued to flower in new ways.

With the coming of the Bolsheviks, however, social "selection pressures" shifted. As part of the Bolshevik campaign to increase scientific literacy, spread Marxist ideology, and undermine religion, Darwinism was systematically popularized in books, pamphlets, and brochures. A natural history collection in Moscow reopened in 1922 as the Museum of Darwinism. Many popular scientific books appeared as part of a new series, edited by M. M. Zavadovsky, called "The Darwinian Library." Plant physiologist Kliment Timiriazev—Darwin's Russian bulldog—lived just long enough to support the Revolution; after his death in early 1920, he became a popular icon, and his many polemics for Darwinism were widely republished. A new 8-volume set of Darwin's complete works (1925–1929) was issued under the editorship of Menzbier, who was still active. Severtsov spent an exceedingly productive decade, capped by his classic work, *Morphologische Gesetzmässigkeiten der Evolution* (1931), which won a hefty state prize. As a result of the new official favor, then, traditional Darwinian work flourished in the postrevolutionary decade.

On the other hand, anti-Darwinian works sometimes ran into trouble. Much controversy surrounded *Nomogenez* (1922) by Leo Berg, a Petrograd geographer and ichthyologist who argued that evolution was governed not by the chance variations of Darwin, but rather by regular, lawful patterns of natural variation whose origin was yet to be understood. As a result of its apparently anti-Darwinian character, the book was almost stopped by the censors (Muller 1925). In the years after its publication Berg's theory acquired quite a following among young Russian biologists (including Dobzhansky), but it was also castigated as finalism, idealism, and mysticism by Marxists and others in a slew of polemical articles, books, and brochures. One was entitled *The Final Word of Anti-Darwinism,* a phrase intended in all three senses (Kozo-Polianskii 1923).[1]

Petrograd's leading experimental biologist also ran into difficulty for challenging Darwin. Trained in Germany under Richard Hertwig, Iurii Filipchenko emerged in the 1920s as Russia's foremost geneticist. Like his Western colleagues, he thought that the field's successful emergence as an experimental science depended on its non-involvement in messy philosophical matters like evolution. Filipchenko regarded evolutionary theory as speculative—but this, of course, did not stop him from speculating. He wrote an impressive history of the evolutionary idea in biology (1923) that demonstrated his mastery of the literature and the issues. Convinced that neither selection nor mutation could adequately explain evolution, he personally favored *autogenesis,* the view that the evolution of a species, like the development of an embryo, was governed by internal causes (perhaps cellular, cytological, or physiological mechanisms). In an influential work published in German (Philiptschenko [Filipchenko] 1927b), he distinguished between intraspecific *microevolution,* which could be studied experimentally, and *macroevolution* above the species level, which he regarded as beyond genetic analysis. By the late 1920s, however, Filipchenko's view had been harshly criticized by Marxists as anti-Darwinian.

Among other experimental biologists, a new trend had developed by mid-decade called "experimental evolution." One of its leading advocates was Kol'tsov, who sought to integrate the experimental cell biology he had learned

[1]The Russian title is *Poslednee slovo anti-darvinizma,* which can mean, loosely translated, "the *final* word of anti-Darwinism" (referring to Berg's philosophical finalism); "the *last* word in anti-Darwinism" (in the sense of "fad"); or "the *last* word of anti-Darwinism" (in the sense of "last gasp").

on the Mediterranean with the classical Darwinism he had learned in Moscow. He is now remembered chiefly for his work on the chemical structure of the gene (1928, 1936), but he also wrote suggestive pieces on biochemical evolution (1929, 1933). His work provides but one example of the Russian interest in biochemical evolution in the 1920s; it is roughly contemporary with that of A. I. Oparin, renowned for his coacervate droplet theory of the chemical origin of life (1924, 1953), and with the work of the brilliant biogeochemist V. I. Vernadsky, who developed the concept of the biosphere (1926). In the early 1920s, Kol'tsov's Institute of Experimental Biology became a center of the new evolutionary trend.

Kol'tsov's most enduring contribution to experimental evolution was his patronage of population genetics. In 1921 he invited a classical entomologist, Sergei Chetverikov, to join his institute in order to preside over the study of an insect that had just recently proved an exemplary experimental object—*Drosophila melanogaster.* Chetverikov had to learn his Morganist genetics from scratch, so he had not learned (as had Filipchenko) that geneticists should not get involved in evolution: as a field naturalist thrust into a laboratory setting, he was curious about what genetics could contribute to traditional evolutionary theory. The result was his classic work, "On Certain Aspects of the Evolutionary Process from the Standpoint of Modern Genetics" (1926), now regarded as one of the earliest statements of the synthetic theory of evolution. In order to test his ideas, Chetverikov and his students began to analyze the genetics of wild populations of drosophila from the environs of Moscow and Dagestan (1925–1926). I have detailed these events elsewhere (Adams 1968, 1970, 1980b, 1989); suffice it to say that these studies of experimental evolution began drosophila population genetics, a field that Russians and their students would dominate for many decades.

Another important experimental evolutionist supported by Kol'tsov was his student Alexander Serebrovsky. Remarkably like H. J. Muller (and only fourteen months his junior), Serebrovsky was an enthusiast for Darwinism, genetics, eugenics, and Marxism who did important research on gene structure, mutagenesis, and human heredity. Unlike Muller, however, Serebrovsky was also a poultry geneticist, and carried out surveys of the domesticated fowl of the tribes of remote mountainous regions in order to find breeding material suitable for selection. These studies led him to develop many terms and techniques central to population genetics, including *gene pool (genofond)* and *gene geography* (Adams 1979). Serebrovsky's microevolutionary work was remarkably innovative, and it influenced Dobzhansky, Timoféeff-Ressovsky, and Dubinin.

Even more ambitious efforts along the same lines were launched by Nikolai Vavilov in *plant science* (a contemporary term used by those who wished to distinguish themselves from old-fashioned botany). An alumnus of the Moscow Agricultural Academy, Vavilov gained fame in 1922 for his law of homologous series of plant variations and was soon hailed as the "Mendeleev of biology," winning political backing from Lenin for an ambitious effort to transform Soviet agriculture through experimental research (Adams 1978). In 1925 he took over buildings at the tsar's summer palace at Detskoe Selo, outside of Leningrad, and started to build up the old Imperial Bureau of Applied Plant Breeding into a new Institute of Applied Botany. To head its laboratory of cytogenetics, he appointed G. A. Levitsky from Kiev, a major center of plant cytology in the heart of Russia's Ukrainian breadbasket. A student of S. G. Navashin (discoverer of double fertilization in plants), Levitsky was the first to use the words *karyotype* and *idiogram* in their modern meanings, and authored the first Russian textbook of cytogenetics (Levitsky 1924 [1976], Adams 1990d). Vavilov also invited a young student from his alma mater, G. D. Karpechenko, who would soon win international fame for his creation of a new species: the true-breeding intergeneric polyploid hybrid of the cabbage and the radish, *Raphanobrassica* (Karpechenko 1928 [1971]; Adams 1990c). In subsequent years, all three would publish widely on plant hybridization, the origin of domesticated plants, and other microevolutionary topics.

By the late 1920s, experimental evolution had gained a following among a young generation of Soviet ecologists. As part of his attempt to create the phytosociology of plant communities (1928b), Petrograd botanist V. N. Sukachev developed ways to measure intraspecific competition and the struggle for existence; his elegant experimental study of the relative viability of northern, central, and southern varieties of dandelions grown together in experimental plots was widely cited in contemporary Western literature as an experimental proof of Darwinian natural selection (e.g., Haldane 1932). As for animals, Moscow biometrician V. Alpatov studied with Raymond Pearl in Baltimore in the late 1920s and returned to develop ecology at Moscow University. His most brilliant pupil was G. F. Gause, whose classic book *The Struggle for Existence* (1934), published in Baltimore under Pearl's auspices, is the locus classicus of the competitive exclusion principle, long a mainstay of ecological theory.

During the postrevolutionary decade, Darwinism was much discussed in philosophical circles, and such leading political figures as Bukharin and Lunacharsky took part. The relationship between Marxism and Darwinism was hotly debated, as were the relative biological and philosophical merits of Darwinism, neo-Darwinism, Lamarckism, neo-Lamarckism, Weismannism, mutationism, vitalism, materialism, and mechanism. In these settings Lamarckism enjoyed special appeal, and Viennese socialist Paul Kammerer accepted an invitation to head a laboratory at the Communist Academy. His suicide weeks later in the face of scandal led to the Soviet film *Salamandr,* whose scenario—written by Lunacharsky, the Bolshevik Commissar of Education—mythologized Kammerer as a revolutionary driven to his end by the invidious machinations of Western capital (Koestler 1971; Gaissinovitch 1980, 1988).

Despite the hardships, then, the postrevolutionary years were filled with achievement and promise for Soviet evolutionists. Because of World War I and the civil war, Russia had been isolated from Western biology for seven years. The evolutionary research she produced in the next decade in a host of fields—morphology, population genetics, biochemical evolution, plant science, ecology—was among the best in world, and in some areas was well ahead of work elsewhere. In evolutionary theory, too, Russia produced some of the decade's best general statements of the old Darwinism (Menzbier and Severtsov), anti-Darwinism (Berg and Filipchenko), and the new Darwinism that was about to be born (Chetverikov and Serebrovsky).

The world knew and appreciated these remarkable contributions because this Russian science was open to the world. Berg's *Nomogenesis* (1926) was published in London only four years after it had appeared in Petrograd. Filipchenko's *Variabilität und Variation* (1927b) and Severtsov's *Gesetzmässigkeiten der Evolution* (1931) were published abroad before they appeared at home (an act that would become illegal from the mid-1930s through the mid-1980s). The Russian achievements of the 1920s would continue to influence the development of evolutionary work in the West in the 1930s, 1940s, and 1950s through the work of Russian émigré biologists. Based in Berlin from 1925 through 1945, Timoféeff-Ressovsky brought the work of the Moscow school to German, Italian, and English readers. Filipchenko's terms *microevolution* and *macroevolution* were introduced into the English evolutionary lexicon by his protégé Theodosius Dobzhansky (1937), from whom they were picked up by Richard Goldschmidt, G. G. Simpson, and many others; likewise Serebrovsky's *genofond,* which Dobzhansky introduced into English as *gene pool* (1950).

At the time, the twenties seemed only the harbinger of more and better science to come; in retrospect, Soviets would look back upon it as the golden age. The many Russian contributions to the development of Western evolutionary thought date almost entirely from this decade; our ignorance of subsequent work is a consequence of what followed.

ISOLATION AND DIVERGENCE: THE STRUGGLE FOR EXISTENCE

Between 1860 and 1930, then, Russian work constituted an especially vigorous geographical variety of Western evolutionary biology. But extreme isolation, accompanied by extraordinarily severe social and ideological selection pressures, soon initiated scientific struggle, divergence, and speciation. Comparable events in Germany would result in a short-lived form of Aryan biology, but while German isolation lasted no more than a decade, Soviet isolation was to last a half-century.

Stalinism. The atmosphere changed radically around 1930. The period of the first five-year plan is known as the "Great Break": It encompassed the consolidation of Stalin's power, the liquidation of the kulaks, widespread ideological crackdowns and institutional reorganizations, and the first show trials. Biological teaching and research emerged from the period restaffed and restructured. Some major figures, including Severtsov, Kol'tsov, and Filipchenko, were driven out of university teaching. University departments and institutes were reorganized and their cadres reconstituted. Foreign contacts diminished precipitately. A Party line was established in most fields and began to permeate scientific debate.

Like most other fields, evolutionary research underwent disruption during the Great Break. At the Kol'tsov Institute, where population genetics had been virtually created, Chetverikov was arrested and exiled and his group disbanded and dispersed. The new ideological line also came to bear on evolutionary theory, where the range of ideologically acceptable opinion narrowed considerably. Filipchenko was posthumously castigated as an idealist for criticizing Darwinism and advocating autogenesis. The position with regard to Lamarckism vacillated: attacked in 1929 and 1930 as a form of unacceptable thinking termed *mechanistic materialism,* Lamarckian viewpoints nonetheless survived in various research settings. Nomogenesis, which had enjoyed considerable popularity in the early 1920s, was now utterly unacceptable on philosophical grounds.

But the imposition of this new isolation was itself to have an important influence on Western evolutionary theory through the founder principle. Largely as a result of the Great Break, two important young Russian biologists failed to return home from abroad as scheduled. Theodosius Dobzhansky had been Filipchenko's protégé in Leningrad beginning in 1924. In December 1927 he came to New York to work in T. H. Morgan's genetics laboratory. As a direct result of the events associated with the Great Break, he decided to stay in America, and a decade later he published *Genetics and the Origin of Species,* a book that triggered the evolutionary synthesis in the United States (Dobzhansky 1937). From the Kol'tsov institute in Moscow, Timoféeff-Ressovsky and his wife had gone to work at the Kaiser Wilhelm Institute for Brain Research in Berlin-Buch in 1925. He remained there until 1945 as one of the Continent's best evolutionary theorists. Indeed, in a recent article, Ernst Mayr has pointed out that "Timofeeff apparently played the same role in Germany [and Italy] that Dobzhansky played in the United States," concluding that "Owing to Timofeeff's influence, an evolutionary synthesis took place in the 1930s in Germany, largely independent of the synthesis in the English-speaking countries" (Mayr 1988: 549). If Einstein was Hitler's gift to American physics, then Dobzhansky and Timoféeff-Ressovsky were Stalin's gift to Western evolutionary biology.

Back in Soviet Russia, largely as a result of the ideological developments during the Great Break, *darvinizm* was enthroned, and the term began to take on a special Soviet flavor. The Darwin celebration of 1932, commemorating fifty years since his death, was widely observed in the Soviet Union in meetings, publications, displays, and the press. In the associated propaganda, Darwinism became identified with a materialist view of life that had been highly praised by Marx and Engels—with some important reservations, notably Darwin's reliance on Malthus and the principle of overpopulation. In addition, Darwinism became linked with the practical task of Soviet rural reconstruction: Darwin's use of examples from practical breeders was emphasized, and it was claimed that, in the Soviet Union, Darwinism would lead to agricultural triumphs and become a principal tool in the Stalinist campaign for the transformation of nature.

The thirties witnessed the consolidation of Stalinism, the centralization of power, the creation and expansion of massive bureaucratic structures, ideological turbulence, and the purges, which swept up between ten and thirty million people. With the imposition of planning and heavy industrialization came parallel ideological campaigns, including Stakhanovism, named after a worker who ostensibly overproduced his production quota through willpower, work, and ideological inspiration. Understandably, it was also a period of intense struggle in Soviet biology.

Lysenkoism. Agriculture witnessed its own Stakhanov in the form of Trofim Lysenko, a Ukrainian peasant who gradually rose to control Soviet biology. He first gained notoriety in 1929 for his technique of vernalization, according to which the heredity of wheat could be changed by the treatment of seeds with suitable regimes of moisture and cold. The result, he claimed, would permit designed alterations in the growing and germination periods of plants, and, ultimately, extra crops. In 1935 he teamed up with I. I. Prezent, his chief ideological theoretician, and the two generalized his vernalization techniques into a new theory of plant physiology. According to this phase theory, during certain periods in its life-cycle a plant's heredity becomes plastic, and relatively simple environmental manipulation could alter the plant's hereditary make-up in quite particular and desirable ways. Soon they generalized their views further into a broadside against the bourgeois, idealist science that opposed such a view of heredity—genetics.

The battle over genetics between Lysenkoists and Vavilovites erupted in 1935, leading to the strident attacks on genetics at the Lenin All-Union Academy of Agricultural Sciences (VASKhNIL) in late 1936, the postponement and then cancellation of the VIIth International Genetics Congress meeting scheduled for Moscow in 1937, the replacement of Vavilov by Lysenko in his agricultural and then in his scientific posts, and the arrest and execution of a number of prominent geneticists, and eventually of Vavilov himself. To some degree, however, Soviet scientists were able to adapt to the new conditions. Despite Chetverikov's earlier arrest and exile, the period saw the continued development and expansion of population genetics, which Kol'tsov was able to reconstitute by inviting to his institute N. P. Dubinin. Together with D. D. Romashov, during the 1930s he headed the institute's so-called "Evolutionary Brigade" of scientific shockworkers. Despite their politically adaptive coloration, they produced many important and influential studies widely quoted in contemporary Western literature.

Much has been written about the history of Soviet genetics and Lysenkoism (e.g., Joravsky 1970, Medvedev 1969). However, most Westerners are not aware of the central role Darwinism played in the political debates. Both the Lysenkoists and their geneticist opponents claimed to be true Darwinists. By then, Darwinism had become a source of legitimacy and, hence, contention.

The struggle for Darwinism. In 1935–1936 Michurin, Pavlov, and Severtsov all died (of natural causes), and all underwent the kind of posthumous canonization so characteristic of Stalin's cult of personality. Following his death in 1935, Ivan Michurin—aptly characterized as the Russian Luther Burbank—began to be portrayed as a great successor to Darwin who extended Darwinism into socialist agriculture and transformed it into a tool for reshaping nature (this, despite the fact that his collected works, issued after 1948 in four volumes under the editorship of Lysenko and Prezent, fail to contain even a single reference to Darwin). Michurin's legacy seemed firmly under the control of Lysenko and Prezent, and they began to call their movement "Michurinist" biology. Meanwhile, shortly before Severtsov's death in December 1936, he was replaced as head of the Institute of Evolutionary Morphology by his prize pupil, Ivan I. Schmalhausen. As Severtsov's chosen successor, Schmalhausen was the executor of Severtsov's legacy and invoked it in the struggle with Lysenko over Darwinism. He was elected to full membership in the USSR Academy of Sciences in 1935 in anticipation of his ascension. After Severtsov's death in late 1936, Schmalhausen moved from Kiev to Moscow, where he took over his teacher's institute and undertook an impressive series of works synthesizing morphology, embryology, population genetics, biogeography, and ecology.

The eightieth anniversary of the publication of the *Origin* occasioned the first skirmishes in the battle for Darwin's mantle, and it saw a remarkable development without parallel in the West: in 1939 Darwinism was declared to be a scientific discipline in its own right, and departments of Darwinism were established at most major universities. Schmalhausen assumed the chair at Moscow University, Prezent at Leningrad. In 1938, in anticipation of the Academy elections of January 1939, Berg was vilified in the press for his earlier opposition to Darwinism; the ground was cleared for Lysenko's direct election as full member of the USSR Academy of Sciences. In August 1940 Vavilov was arrested. As the case against him was being prepared, some of his closest associates followed him to prison, among them two of the world's most renowned plant cytogeneticists, G. D. Karpechenko and G. A. Levitsky. All three died in prison in 1942 or 1943.

During World War II there was an understandable lull in these struggles over Darwin's legacy, but Russia's resilient scientific community was able to make major advances in evolutionary theory. Scientific personnel from Leningrad and Moscow were evacuated to Central Asia and other remote regions where they continued their work. When the Red Army reached Berlin, it made Timoféeff-Ressovsky director of his Kaiser Wilhelm Institute. Shortly thereafter, however, he was arrested by the NKVD and imprisoned, remaining in secret police prison research camps (*sharashki*) until the mid-1950s. But the Soviet Union's other two leading evolutionary theorists prospered with the war's end. In 1946, Dubinin was elected a corresponding member of the Academy largely for his work on population genetics and evolutionary theory. Schmalhausen published two outstanding products of his wartime labor: His theoretical classic *Faktory evoliutsii* (1946a), translated into English by Dobzhansky in 1949 as *Factors of Evolution,* and his important synthetic text *Problemy darvinizma* [Problems of Darwinism] (1946b). This work, together with that by A. A. Paramonov, *Kurs darvinizma* (1945), were approved by the appropriate ministries as the first Soviet textbooks in the new science of Darwinism. They compare favorably to contemporary American and European textbooks on evolution.

In the three years following the war, the struggle over Darwinism intensified. At its core was a debate over the reality and significance of intraspecific competition. In a series of articles, Lysenko claimed that overpopulation was rare in nature and played no evolutionary role. Rather, he held that natural selection and the struggle for existence operated not between members of the same species but between members of different species: Not intraspecific competition, but interspecific competition was key. Citing lengthy evidence from experimental plots, Lysenko claimed that the plant most often used (by Sukachev, Schmalhausen, and others) to demonstrate intraspecific struggle—the Russian dandelion (Kok-sagyz)—could be shown actually to increase its biomass and prosper when crowded in with con-specific, closely allied varieties.

To classical Darwinists East and West, this position represented a direct attack on the Darwinian core. Schmalhausen organized a national conference at Moscow University called "Problems of Darwinism." Among the participants was Iurii Zhdanov, head of the Central Committee's science section, who supported Schmalhausen's position and rejected Lysenko's. The meeting was widely regarded as an attack on Lysenko's evolutionary revisions, and with some justice, but its proceedings were never published. Lysenko's creative Darwinism had found an ally in Stalin.

"CREATIVE DARWINISM"

> Comrades, before I pass to my concluding remarks I consider it my duty to make the following statement.
> The question is asked in one of the notes handed to me, What is the attitude of the Central Committee of the Party to my report? I answer: The Central Committee of the Party examined my report and approved it. (*Stormy applause. Ovation. All rise.*)
> . . . *Glory to the great friend and protagonist of science, our leader and teacher, Comrade Stalin! (All rise. Prolonged applause.)*
>
> **T. D. Lysenko, 7 August 1948**

At the infamous August 1948 session of the Lenin All-Union Academy of Agricultural Sciences, Lysenko's biological views became state policy. We associate the meeting with the destruction of Soviet genetics, the "death of a science in Russia" (Zirkle 1949). The observation is not quite accurate, but it is fair. On the other hand, some have also characterized the meeting as the repudiation of Darwin's theory of evolution. This well-meaning generalization is fundamentally misleading, and it hides the most historically informative aspects of the event.

August 1948. The central presentation of the August meeting was Lysenko's presidential address, "The Situation in Biological Science." Lysenko began and concluded his address by discussing Darwinism, and his indictment of geneticists in the intervening pages centered on his contention that they had perverted and misunderstood its essence (Lysenko 1948: 11–50). "The appearance of Darwin's teaching," Lysenko began, expounded in *The Origin of Species,* "marked the beginning of scientific biology." Its central idea of "selection"—"scientific and true"—was in essence but "a summation of the age-old practical experience of plant and animal breeders who, long before Darwin, produced varieties of plants and breeds of animals by the empirical method" (11–12).

Of course, "although unquestionably materialist in its main features," Darwin's theory was not perfect: he made mistakes, foremost among them his use of reactionary Malthusian ideas. "It must be clear to any progressively thinking Darwinist," observes Lysenko, "that, even though Darwin accepted Malthus' reactionary theory, it basically contradicts the materialist foundation of his own teaching." But apparently the great Darwin sensed the problem, because "under the weight of the vast amount of biological facts accumulated by him, Darwin felt constrained in a number of cases radically to alter the concept of the 'struggle for existence,' to stretch it to the point of declaring that it was just a figure of speech" (13–14).

Alas, Darwin's legacy had been misunderstood and perverted. "In the post-Darwinian period the overwhelming majority of biologists—far from further developing Darwin's teaching—did all they could to debase Darwinism, to smother its scientific foundation." And, of course, "the most glaring manifestation of such debasement of Darwinism is to be found in the teachings of Weismann, Mendel, and Morgan, the founders of modern reactionary genetics." "Reactionary biologists have therefore done everything in their power to empty Darwinism of its materialist elements," and they have done so by inadmissibly "representing these erroneous aspects of Darwinian theory, those based on Malthus' theory of overpopulation with the inference of a struggle presumably going on within species," as the "cornerstone of Darwinism" (14–15).

In sum, then, according to Lysenko, "The Mendelist-Morganists cling to everything that is obsolete and wrong in Darwin's teaching, at the same time discarding its living materialist core" (27). Not so the Michurinists. Taking "the Darwinian theory of evolution as their basis," they have "transformed it in the light of the teachings of Michurin and Williams and thereby converted [it] into Soviet creative Darwinism." Their achievement, in Lysenko's view, is that by following the correct essence of Darwin's theory, they have transformed Darwinism "from a science which primarily *explains* the past history of the organic world" into "a creative, *effective means* of systematically mastering living nature, making it serve practical requirements," making it "capable of helping the tiller of the soil to obtain two ears of corn where there was formerly only one" (14, 46–47). Lysenko is explicit about the tenets of the new Darwinism, the following among them: (1) *"Heredity is the effect of the concentration of the action of environmental conditions assimilated by the organism in a series of preceding generations"*; (2) "It is possible . . . to *force* any form of animal or plant *to change more quickly and in a direction desirable to man"*; (3) "there exist no intraspecific struggle and mutual assistance among individuals within a species"; (4) "there does exist interspecific struggle and competition and also mutual assistance between different species"; (5) *"The conversion of one species into another takes place by a leap"*; in particular, a member of one existing species can be transformed into a member of a totally different current species, even of a different genus, in a single generation or a considerably shorter time.

Lysenko reserved his heaviest attack for those evolutionary theorists whose work was linked with genetics, notably Schmalhausen and Dubinin. In the aftermath of the meeting, Schmalhausen and Dubinin (as well as many others) lost their jobs, and Lysenko's brand of Darwinism became the officially sanctioned one. In an interview in 1953, Lysenko claimed that his text had been edited and approved by Stalin personally.[2] During the period 1948–52,

[2]There now appears to be documentary proof of his claim: a copy of a transcript of Lysenko's speech, with Stalin's editing, has recently been located in the VASKhNIL archives by Soviet historian of science, Kiril Rossiyanov.

Lysenko extended his revision of Darwin to the question of the origin of species, claiming that Avakian and other workers had succeeded in transforming a hawkweed into a hazelnut in one generation. For four years—from late 1948 until late 1952—these views became part of officially sanctioned "Creative Darwinism" (*tvorcheskii darvinizm*).

The other synthesis. In some respects Lysenko's "theory" (which I believe owed much to his philosophical collaborator I. I. Prezent) represented an attempt to create an alternate evolutionary synthesis, one that deploys the Darwinian legacy in what is, from our perspective, a novel way. A comparison between the views of Schmalhausen and Lysenko highlights the similarities and differences between the two approaches.

In simplified terms, both Lysenko and Schmalhausen agreed that Darwin was great and Darwinism good, that Darwin relied on the work of practical breeders, that he believed in correlated variability and the inheritance of acquired characteristics, that he asserted the importance of overpopulation in causing a natural selection favoring adaptive variants. Both agreed that the core of Darwinism was essentially correct, that Darwin had been mistaken in certain respects, and that modern Darwinism was advancing by quite properly correcting his errors.

They differed, however, in their interpretation of *what constituted* the true core of Darwin's thought. Like most Western creators of the evolutionary synthesis, Schmalhausen believed that the essence of Darwinism was the natural selection of intraspecific hereditary variants to produce evolutionary change. Where Darwin had been mistaken was in his view of inheritance and of the origin and nature of these variants, and most especially in his adherence to the inheritance of acquired characteristics: He had been too credulous in believing the practical knowledge of the breeders. Indeed, Darwin had openly admitted his failing in the phrase, "We are profoundly ignorant of the causes of variation." This weakness in Darwinism, then, had been corrected by the development of modern genetics.

Lysenko held the converse view. For him, Darwin's attitude toward heredity and his reliance of the knowledge of practical breeders were correct, as was his view that organisms evolved and adapted as the result of natural, "material" processes dependent on the environment. But Lysenko held that he had erred in applying Malthusianism to biology and in relying too heavily (because of his class background and capitalist British setting) on intraspecific competition and overpopulation. That was the mistake which Lysenko's Michurinist theoretical advance was correcting. And Lysenko emphasized that whereas the other Darwinism made mankind largely powerless in transforming nature (having to rely on the selection of variants as they randomly appear), his own creative Darwinism extended and expanded the breeders' knowledge so as to mold organic forms, allowing Darwinism to be mobilized in agriculture.

Meanwhile, Lysenko's theories on the absence of intraspecific competition in nature found direct application in the Forest Belt Campaign. Wide-scale tracts devastated by the war were reforested according to Lysenko's approach—not by spacing, but by clustering the seedlings in what came to be called the "cluster method." By mid-1952, it was evident to many that Lysenko's cluster method had proved catastrophic, leading to a loss of some 80% of the seedlings. To no one was this more evident than the "grand old man" of Russian botany, V. N. Sukachev.

A member of both the Communist Party and the Academy of Sciences (1937/1943), the founder and director of its Forestry Institute, and editor of the prestigious *Botanical Journal,* Sukachev was strategically placed to move against the Michurinist hegemony. In late 1952 (while Stalin was still alive), Sukachev launched a widespread discussion of Darwinism, the species theories, and intraspecific competition. The opening salvo consisted of two articles in the December issue criticizing Lysenko's revision of Darwinism, arguing that his views differed greatly from Darwin's and that evidence from both science and Marxism established that, between the two, Darwin's was the correct view. The discussion was joined in articles by Sukachev himself and many others, together with occasional outraged Lysenkoist rejoinders, and these meaty and elaborate expositions, printed in a small font, filled the journal's discussion section for six years. The effect was to uncouple Darwinism from Lysenko's theories for purposes of scientific discussion in technical journals.

However, from 1949 until 1964, the public, popular, and pedagogical image of Darwinism remained firmly under Lysenkoist control. During this period, all the Darwinism textbooks used in Soviet schools and universities were written by Lysenkoists (Alekseev, Dvoriankin, Prezent, Veselov) and they demonstrate an alternative way of deploying Darwin's scientific legacy. In comparison to the earlier texts of Paramonov and Schmalhausen, these Michurinist texts leave the account of the history of life, systematics, comparative anatomy, paleontology, and similar subjects largely intact. In place of the genetics and population genetics sections, however, they include lengthy treatments of Michurin, Lysenko, and Michurinism, depicting such work as applied, practical Darwinism.

The Darwin Centennial of 1959 coincided with a rise in Lysenko's fortunes. In a vulnerable position following Stalin's death, Lysenko won Khrushchev's support in late 1958 by his personal loyalty in the leadership struggles which climaxed in that year; one result was Sukachev's removal as editor of *Botanical Journal* and the replacement of almost its entire editorial board. Lysenko's new confidence is evident in his centennial anthology, *Darwinism Lives and Develops,* in which he once again advocated his theories of evolution. Most of Russia's best evolutionary theorists adopted a low profile during this Lysenkoist period, working on safe topics like radiation genetics, DNA chemistry, or comparative anatomy. Meanwhile, powerful physical scientists in the Academy were working to free Soviet biological research from Lysenko's administrative control. In 1963, largely in order to remove molecular biology from Lysenko's control, the Biology Division of the Academy was split into three new divisions (physiology, molecular biology, and general biology): the tactic worked but, unfortunately, evolutionary biology was left in the same division as Lysenko and his followers, a fact whose consequences for evolutionary biology can still be felt today.

PROSPECTS UNDER PERESTROIKA

In the little October revolution of 1964, Khrushchev was ousted from power, and shortly thereafter Lysenko's hegemony ended. In the aftermath, the Academy reexamined and largely repudiated Lysenko's biology. Serious attempts were made to reconstitute Soviet Darwinism, but the repudiation of Michurinism in evolutionary theory was not so rapid or widespread as in genetics and molecular biology. Multi-authored works were issued from Leningrad University on the history of *evoliutsionnoe uchenie* [evolutionary science] (1966) and current evolutionary theory (1967); a series of annual volumes *(Problemy evoliutsii)* began to appear in 1968 under the editorship of N. N. Vorontsov; up-dated versions were issued of the original Darwinism texts of Schmalhausen (1969) and Paramonov (1978); and new works on evolutionary theory by Timoféeff-Ressovsky (1969) and Iablokov (1969, 1976) were published. Of course, Davitashvili issued new books, and new editions appeared of former Lysenkoist texts (e.g., that of Pravdin: 1960, 1964, 1968), but they encountered heavy criticism in lengthy reviews. Meanwhile, a new generation of leaders in evolutionary studies emerged: Takhtadzhan (paleobotany), Vorontsov, Iablokov, and others, many of whom count themselves as students of Schmalhausen or Timoféeff-Ressovsky.

With these developments has also come a greatly diminished use of the term *darvinizm* and its history has been revised. These historical developments are reflected in the new Soviet historiography of Darwinism. The new view is more international in scope, emphasizing the important (but previously unrecognized) contributions of Soviet biologists to the world history of evolutionary theory. Works by Serebrovsky, Chetverikov, Vavilov, Karpechenko, and even Leo Berg have been published or reissued posthumously, and the history of evolutionary thought in the Soviet Union and the West has been discussed in first-rate historical monographs. At the same time, a half-century of ideology and isolation have left their mark. As in former days, the literature on the history of Russian science remains obsessed with the heroic personalities and scientific brilliance of Soviets and with some remarkable claims about their scientific priority. The principal difference between the old literature and the new seems to be that in Stalinist days, their failure to be recognized was blamed on the bourgeois, capitalist West, whereas today it is blamed on homegrown Stalinist repression.

These works are interesting to compare with the products of our own Darwin industry, which has concentrated considerable effort and ingenuity on the background, writings, and thought of Darwin himself. By contrast, only one Soviet book on Darwin has appeared in recent years; instead, Soviet workers have lavished their attention on what may be justly called the "history of Darwinism," and Soviet works on the post-Darwin history of evolutionary theory and thought outnumber Western works on the subject two to one. It is not difficult to understand why. Until the 1960s the historiography of Soviet Darwinism had been linked with Lysenkoism. With Lysenko dethroned and classical genetics once again legitimate, a new historiography came into being. Our Western conceptions of Darwin and Darwinism—the myths that our scholarship seeks to dispel as well as the legitimate heirs it seeks to crown—are all products of the history of Darwinism in America and Britain. Soviet scholarship is no less a product of Soviet Darwinism, for it has had its own myths to dispel, and other legitimate heirs to crown.

Since 1986, with the consolidation of power by Gorbachev, the Soviet Union has experienced a blossoming of open expression which the world has come to know as *glasnost'*. One of its consequences has been greatly increased contact between Soviet and Western scientists. In some respects, on some levels, Soviet isolation appears to be ending. But the effects of fifty years of isolation cannot be easily brushed aside. If the process of scientific speciation has ceased, Soviet evolutionary biology nonetheless constitutes a strongly pronounced variety rather more distinct than before, and its reintegration into world science poses both opportunities and challenges.

Structurally, Soviet science seems virtually washed clean of the more extreme forms of Marxist-Leninist rhetoric and ideology. If current trends continue, it appears as though the discussion that has so exercised Western analysts for so many years—the relationship between science and dialectical materialism in the Soviet Union—may cease to have a Russian referent. But if ideology has disappeared, it has left behind a massive state bureaucratic structure, fed for years by privileged access to resources, highly uneven in the education and quality of its personnel, and technologically handicapped. It remains big science, Soviet-style: State-sponsored, centralized, bureaucratic, nationalist, and isolated. With international English as the new *lingua franca* of science, current Russians seem less adept at foreign languages than any generation of Russian scientists in history. And astonishing challenges face Russian evolutionary biologists on the social front. Chernobyl may be just the tip of the iceberg: many regions of the Soviet Union are ecological disaster areas. Even with the promising rise to political prominence of such leading Soviet evolutionary biologists as Vorontsov (new minister of the environment) and Iablokov (in the parliament), it remains to be seen what can be done.

But this troubling picture has a brighter side, and the West may have to gain much from renewed contacts. The best Soviet scientists are part of a brilliant, sophisticated, astonishingly cultivated intelligentsia that has largely disappeared in much of Western science. If anything, the years of repression seem to have intensified its commitment to its own pre-Stalinist traditions and legacies. In evolutionary studies, Soviets have continued to develop the traditions of classic physiology, morphology, and embryology, and these fields influence their evolutionary thinking in a way rarely encountered in Western biology. In addition, the extraordinary brilliance of the Russian mathematical tradition appears to play a rather larger role in Soviet theorizing than in the West. Their isolation has caused Russian biologists to observe many exciting Western developments only from afar. But Western biologists have also missed something: access to the rich natural history and biological traditions—the morphological, mathematical, ecological, and systematical thinking—of which such émigrés as Dobzhansky and Timoféeff gave us only a glimpse.

Russian culture in the 19th century provides a hopeful parallel. After decades of isolation, beginning in 1855 a series of social reforms initiated a brilliant half-century of Russian science and culture. Just as Europe was growing tired of the novel in literature and romanticism in music, Russia invigorated the world with the contributions of Dostoevski and Tolstoi, of Tchaikovsky and Rimsky-Korsakov. And by the turn of the century Russians and Russian émigrés would be breaking new ground in music, dance, art, and literature. Will the brilliant evolutionary traditions of Russia similarly infuse Western evolutionary biology with a fresh vitality, with old ideas become new? I hope that Russian isolation has truly come to an end so that we may all find out; I suspect that Western evolutionary biology may have more to gain than it imagines.

REFERENCES

Adams MB (1968): The founding of population: Contributions of the Chetverikov School, 1924–34." J Hist Biol 1(1):23–39.

Adams MB (1970): Towards a synthesis: Population concepts in Russian evolutionary thought, 1925–35. J Hist Biol 3(1):107–29.

Adams MB (1977/78): Biology after Stalin: A case study. Survey: J East/West Stud 23(1):53–80.

Adams MB (1977/78): Vavilov, Nikolay Ivanovich. "Dictionary of Scientific Biography," supp. 1(15), S.V. New York: Chas Scribner's Sons.

Adams MB (1979): From "gene fund" to "gene pool": On the evolution of evolutionary language. In Coleman W, Limoges C (eds): "Studies in History of Biology." Baltimore: Johns Hopkins Univ Press, pp 241–85.

Adams MB (1980a): Science, ideology, and structure: The Kol'tsov Institute, 1900–1970. In Lubrano LL, Solomon SG (eds): "The Social Context of Soviet Science." Boulder, CO: Westview Press, pp 173–204.

Adams MB (1980b): Sergei Chetverikov, the Kol'tsov Institute, and the evolutionary synthesis. In Mayr E, Provine WB (eds): "The Evolutionary Synthesis: Perspectives on the Unification of Biology." Cambridge (MA): Harvard Univ Press.

Adams MB (1980c): Severtsov and Schmalhausen: Russian morphology and the evolutionary synthesis. In Mayr E, Provine WB (eds): "The Evolutionary Synthesis: Perspectives on the Unification of Biology." Cambridge (MA): Harvard Univ Press.

Adams MB (1990a): Chetverikov, Sergei Sergeevich. "Dictionary of Scientific Biography," supp. 2 (17), s.v. New York: Chas Scribner's Sons, pp 155–165.
Adams MB (1990b): Filipchenko, Iurii Aleksandrovich. "Dictionary of Scientific Biography," supp. 2 (17), s.v. New York: Chas Scribner's Sons, pp 297–303.
Adams MB (1990c): Karpechenko, Georgii Dmitrievich. "Dictionary of Scientific Biography," supp. 2 (17), s.v. New York: Chas Scribner's Sons, pp 460–464.
Adams MB (1990d): Levitskii, Grigorii Andreevich. "Dictionary of Scientific Biography," supp. 2 (18), s.v. New York: Chas Scribner's Sons, pp 549–553.
Adams MB (1990e): Oparin, Aleksandr Ivanovich. "Dictionary of Scientific Biography," supp. 2 (18), s.v. New York: Chas Scribner's Sons, pp 695–700.
Adams MB (1990f): Serebrovskii, Aleksandr Sergeevich. "Dictionary of Scientific Biography," supp. 2 (18), s.v. New York: Chas Scribner's Sons, pp 803–811.
Astaurov BL, Rokitskii PF (1975): "Nikolai Konstantinovich Kol'tsov." Moscow: Nauka.
Avakian AA, Dolgushin D, Belen'kii N, Glushchenko I, Dvoriankin F (1947): Za darvinizm tvorcheskii, protiv mal'tuzianstva [For creative Darwinism, against Malthusianism]. Literaturnaia gazeta, 29 Nov.
Basalla G (1988): "The Evolution of Technology," Cambridge: Cambridge Univ Press.
Berg LS (1922): "Nomogenez ili Evoliutsiia na Osnove Zakonomernostei" [Nomogenesis, or evolution determined by laws]. Petrograd: Gosudarstvennoe izdatel'stvo.
Berg LS (1926): "Nomogenesis or Evolution Determined by Law." London: Constable.
Berg LS (1977): "Trudy po teorii evoliutsii, 1922–1930" [Works on evolutionary theory, 1922–30]. Leningad: Nauka.
Bowler P (1983): "The Eclipse of Darwinism." Baltimore: Johns Hopkins Univ Press.
Chetverikov SS (1926): O nekotorykh momentakh evoliutsionnogo protsessa s tochi zreniia sovremennoi genetiki [On certain aspects of the evolutionary process from the standpoint of modern genetics.] Zhurnal eksperimental'noi biologii, ser. A, no. 1:3–54.
Chetverikov SS (1928): Über die genetische Beschaffenheit wilder Populationen. Zeitschrift für induktive Abstammungs- und Vererbungslehre 46 (2):1499–1500.
Chetverikov SS (1961): On certain aspects of the evolutionary process from the standpoint of modern genetics (Barker M, trans; Lerner IM, ed). Proc Amer Phil Soc 105(2): 167–95.
Chetverikov SS (1983): "Problemy obshchei biologii i genetiki (vospominaniia, stat'i, lektsii)" [Problems of general biology and genetics (memoirs, articles, lectures)], Nikoro ZS (ed). Novosibirsk: Nauka.
Chetverokov SS: "Darvinizm zhivet i razvivaetsia" [Darwinism lives and develops] (1960): Moscow: Izdatel'stvo Akademii nauk SSSR.
Dobzhansky T (1937): "Genetics and the Origin of Species." New York: Columbia Univ Press.
Dubinin NP, Romashov DD (1932): Geneticheskoe stroenie vida i ego evoliutsiia: 1. Genetiko-avtomaticheskie protsessy i problema ekogenotipov [Genetic structure of the species and its evolution: 1. Automatic genetic processes and the problem of ecogenetypes]. Biologicheskii Zhurnal 5-6:52–95. Dubinin NP [and 14 coauthors] (1934). Eksperimental'nyi analiz ekogenotipov *Drosophila melanogaster* [Esperimental analysis of ecogenotypes of *Drosophila melanogaster*], parts 1 and 2. Biologicheskii Zhurnal 1:166–216.
Dubinin NP, Geptner MA, Demidova ZA, D'iachkova LI (1936). Geneticheskaia struktura populiatsii i ee dinamika v dikikh naseleniiakh *Drosophila melanogaster* [Genetic structure of populations and its dynamics in wild *Drosophila melanogaster*]. Biologicheskii Zhurnal 6:939–76.
Dubinin NP, Romashov DD, Geptner MA, Demidova ZA (1937): Aberrativnyi polimorfizm u *Drosophila fasciata* Meig. (Syn. *D. melanogaster* Meig.) [Aberrant polymorphism in *Drosophila fasciata* Meig.]. Biologicheskii Zhurnal 2:311–54.

Dunn LC (1944): Science in the Soviet Union: I. Genetics. Science and Society 8(1): 64–66.
Dunn LC (1965): "A Short History of Genetics." New York: McGraw–Hill.
Eitingen G (1951): Gnezdovye prosevy lesnykh polos [Cluster planting of forest belts]. Izvestiia, 19 Sept.
Feiginson (1951): "Sovetskii tvorcheskii darvinizm" [Soviet creative Darwinism]. Moscow: Gosudarstvennoe izdatel'stvo kul'turnoprosvetitel'noi literatury.
Filipchenko IA (1915): "Izmenchivost' i evoliutsiia" [Variation and evolution]. Petrograd: Biblioteka naturalista.
Filipchenko IA (1923): "Izmenchivost' i metody ee izucheniia" [Variation and methods for its study]. Petrograd: Gosizdat. [ed 4, 1929, Leningrad: Gosizdat.]
Filipchenko IA (1927a): Uspekhi genetiki za poslednie 10 let (1918–1927) v SSSR [Achievements of genetics in the USSR over the past ten years (1918–27)]. Trudy Leningradskogo obshchestva estestvoispytatelei 62(1):3–11.
Filipchenko IA [Philiptschenko J] (1927b). "Variabilität und Variation." Berlin: Borntraeger.
Filipchenko IA (1977): "Evoliutsionnaia ideia v biologii: istoricheskii obzor evoliutsionnykh uchenii XIX veka" [The evolutionary idea in biology: Historical survey of evolutionary thought in the nineteenth century]. ed 3. Moscow: Nauka.
Gaissinovitch AE (1980): The origins of Soviet genetics and the struggle with Lamarckism, 1922–29. Adams MB, trans. J Hist Biol 13(1):1–51.
Gaissinovitch AE (1988): Zarozhdenie i razvitie genetiki [The origins and development of genetics]. Moscow: Nauka.
Gall IM (1976): "Bor'ba za sushchestvovanie kak faktor evoliutsii (istoriko-kriticheskii analiz otechestvennykh botanicheskikh issledovanii)" [Struggle for existence as a factor of evolution (critical historical analysis of native botanical research)]. Leningrad: Nauka.
Gause GF (1934): "The Struggle for Existence." Baltimore: Johns Hopkins Univ.
Gershenson SM (1934): Mutant genes in a wild population of *Drosophila obscura* Fall. Amer Naturalist 68(719):569–71.
Glushchenko I (1949): Reaktsionnaia genetika na sluzhbe imperializma [Reactionary genetics in the service of imperialism]. Pravda, 5 April.
Graham LR (1967): "The Soviet Academy of Sciences and the Communist Party, 1927–1932." Princeton: Princeton Univ Press.
Graham LR (1972): "Science and Philosophy in the Soviet Union." New York: Alfred A. Knopf.
Greene J (1981): "Science, Ideology, and World View: Essays in the History of Evolutionary Ideas." Berkeley: Univ Calif Press.
Haldane JBS (1932): "The Causes of Evolution." London: Longmans.
Hull DL (1988): "Science as a Process: An Evolutionary Account of the Social and Conceptual Development of Science." Chicago: Univ Chicago Press.
Ivanov ND (1952): O novom uchenii T. D. Lysenko o vide [Concerning T. D. Lysenko's new doctrine on species]. Botanicheskii Zhurnal 6:819–42.
Ivanov ND (1960): Darvinizm i teorii nasledstvennosti" [Darwinism and theories of heredity]. Foreword by Davitashvili LS. Moscow: Izdatel'stvo Akademii nauk SSSR.
Joravsky D (1961): "Soviet Marxism and Natural Science, 1917–1932." Stud Russian Inst, Columbia Univ. New York: Columbia Univ Press.
Joravsky D (1965): The Vavilov brothers. Slavic Rev 24(3):381–94.
Joravsky D (1970): "The Lysenko Affair." Cambridge (MA): Harvard Univ Press.
Kaftanov SV (1948): Za bezrazdel'noe gospodstvo michurinskoi biologicheskoi nauki [For the undivided rule of Michurinist biological science]. Moscow: Pravda.

Kapitsa PL (1962a): Budushchee nauki [The future of science]. Nauka i Zhizn' 3:18–23, 96–97.
Kapitsa PL (1962b): Teoriia, eksperiment, praktika [Theory, experiment, practice]. Ekonomicheskaia gazeta, 26 March.
Kapitsa PL (1974): "Eksperiment, teoriia, praktika: stat'i, vystupleniia" [Experiment, theory, practice: essays and addresses]. Moscow: Nauka.
Kapitsa PL (1980): "Experiment, Theory, Practice: Articles and Addresses." Dodrecht, Holland: D. Reidel Publ Co.
Karpechenko GD (1971): "Izbrannye trudy" [Selected works]. Moscow: Nauka.
Karpechenko GD: "Klassiki sovetskoi genetiki 1920–1940" [Classics of Soviet genetics 1920–40] (1968). Leningrad: Nauka.
Koestler A (1971): "The Case of the Midwife Toad." London: Hutchinson.
Kol'tsov NK (1924): Experimental biology and the work of the Moscow Institute. Science 59(1536):497–502.
Kol'tsov NK (1928): "Fiziko-khimicheskie osnovy morfologii." Zhurnal eksperimental'noi biologii, Seriia B(1):3–31.
Kol'tsov NK (1936): "Organizatsiia kletki: Sbornik eksperimental'nykh issledovanii, statei i rechei 1903–1935 gg." [The organization of the cell: a collection of experimental research papers, articles, and speeches, 1903–35]. Moscow and Leningrad: Gosudarstvennoe izdatel'stvo biologicheskoi i meditsinskoi literatury.
Kol'tsov NK (1939): "Les molécules héréditaires." Actualités scientifiques et industrielles, no. 776: Exposés de génétique (La Génétique et les problèmes de l'évolution, no. 2.) Paris: Hermann and Co.
"Konferentsiia po problemam darvinizma (3–8 fevralia 1948 goda): Tezisy dokladov" [Conference on problems of Darwinism (3–8 February 1948): paper abstracts] (1948): Moscow: Izdanie Moskovskogo gosudarstvennogo universiteta.
Kozo-Polianskii BM (1923): "Poslednee slovo antidarvinizma." Krasnodar.
Kuhn T (1962): "The Structure of Scientific Revolutions," Chicago: Univ Chicago Press.
Lepeshinskaia, OB (1952): "Kletka: ee zhizn'ki proiskhozhdenie" [The cell: its life and origin]. Moscow: Gosudarstvennoe izdatel'stvo meditsinskoi literatury.
Levitskii GA (1976): "Tsitologiia rastenii: Izbrannye trudy" [Plant cytology: Selected works]. Moscow: Nauka.
Lobashev ME (1963): "Genetika: Kurs lektsii" [Genetics: A lecture course]. Leningrad: Izdatel'stvo Leningradskogo universiteta. [ed 2, 1967].
Lobashev ME (1969): "50 let kafedre genetiki i selektsii Leningradskogo universiteta" [Fifty years of the Department of Genetics and Selection at Leningrad University]. Genetika 10:182–89.
"Lysenko, Trofim Denisovich" (1953): Glushchenko IE, intro; Epifanova AP, comp. Akademiia nauk SSR. Materialy k biobibliografii uchenykh SSSR. Seriia biologicheskikh nauk. Agrobiologii no. 1. Moscow: Izdatel'stvo Akademii nauk SSSR.
Lysenko TD (1946): "Heredity and its Variability," Dobzhansky T (trans). New York: King's Crown Press. [Reprinted 1949: Columbia Univ Press.]
Lysenko TD (1948): "Agrobiologiia: Raboty po voprosam genetiki, selektsii, i semenovodstva" [Agrobiology: works on genetics, selection, and seed breeding]. ed 4. Moscow: Gosudarstvennoe izdatel'stvo sel'skokhoziaistvennoi literatury. [ed 6, 1952.]
Lysenko TD (1948): On the situation in biological science. In "The Situation in Biological Science." pp 11–50.
Lysenko TD (1950): Novoe v nauke o biologicheskom vide [What is new in science concerning the biological species]. Pravda, 3 Nov. [Reprinted 1953, Botanicheskii Zhurnal 1:44–56; Bol'shaia sovetskaia entsiklopediia. ed 2. s.v. Vid (Species).]

Lysenko TD (1959): Darvinizm zhivet i razvivaetsia [Darwinism lives and develops]. Izvestiia, 22 Nov. [Reprinted 1961 in Trudy Instituta genetiki 28:24–29.]

Lysenko TD (1960): O zakone zhizni biologicheskikh vidov i ego znachenii dlia praktiki [On the law of the life of biological species and its significance for practice]. In Nasledstvennost' i izmenchivost' 1:212–235.

Mayr E, Provine WB (eds) (1980): "The Evolutionary Synthesis: Perspectives on the Unification of Biology." Cambridge MA: Harvard Univ Press.

Medvedev NN (1978): "Iurii Aleksandrovich Filipchenko, 1882–1930." Moscow: Nauka.

Medvedev ZA (1969): "The Rise and Fall of T. D. Lysenko," Lerner IM (trans). New York: Columbia Univ Press.

Medvedev ZA (1971): "The Medvedev Papers: Fruitful Meetings Between Scientists of the World; Secrecy of Correspondence is Guaranteed by Law," Rich V (trans), foreword by Ziman J. London: Macmillan, St. Martin's Press.

Menzbir MA (ed) (1925–1929): "Polnoe sobranie sochinenii Char'lza Darvina" [Complete collected works of Charles Darwin]. 8 vols. Moscow/Leningrad.

Michurin IV (1948): "Sochineniia v chetyrykh tomakh" [Works in four volumes], Lysenko TD (ed). Vol 1, "Printsipy i Metody Rabot" [Principles and techniques of work]. ed 2. Moscow: Gosudarstvennoe izdatel'stvo sel'skokhoziaistvennoi literatury.

Morgan TH, Sturtevant AH, Muller HJ, Bridges CB (1915): "The Mechanism of Mendelian Heredity." New York: Holt.

Moskovskaia konferentsia po problemam darvinizma [Moscow conference on problems of Darwinism] (1948): Priroda 6:85–87.

"O polozhenii v biologicheskoi nauke. Stenograficheskii otchet sessii Vsesoiuznoi akademii sel'skokhoziaistvennykh nauk imeni V.I. Lenina 31 iiulia-7 avgusta 1948 g." [On the situation in biological science. Stenographic records of the session of the VASKhNIL held 31 July7–August 1948]. Moscow: Gosudarstvennoe izdatel'stvo sel'skokhoziaistvennoi literatury.

O rezul'tatakh proverki deiatel'nosti eksperimental'noi bazy i podsobnogo khoziaistva 'Gorki Leninskie' Akademii nauk SSSR: Sovmestnoe zasedanie prezidiuma Akademii nauk SSSR, kollegii Ministerstva sel'skogo khoziaistva SSSR, i prezidiuma VASKhNIL 2 sentiabria 1965 g. [On the results of the investigation of the Lenin Hills experimental station and farm of the AN SSSR: Joint session of the presidium AN SSSR, the collegium of the Ministry of Agriculture of the USSR, and the presidium of the VASKhNIL, 2 September 1965]: Vestnik Akademii Nauk SSSR 11:1–128.

Ob agrobiologicheskoi nauke i lozhnykh pozitsiiakh Botanicheskogo Zhurnala [Concerning agrobiology and false positions taken by Botanical Journal] (1958): Pravda 14 Dec.

Ob utverzhdenii sostava redaktsionnoi kollegii 'Botanicheskogo zhurnala [On the confirmation of the composition of the editorial board of Botanical Journal] (1959): Vestnik Akademii Nauk SSSR 3:112.

Oparin AI (1924): "Proiskhozhdenie zhizni" [The Origin of Life]. Moscow: Izdatel'stvo "Moskovskii rabochii."

Oparin AI (1953): "The Origin of Life," Morgulis S (trans). ed 2. New York: Dover Publ.

Paramonov AA (1945): "Kurs darvinizma" [Course in Darwinism]. Moscow: Sovetskaia nauka. [ed 2, 1978.]

Philiptschenko J. See Filipchenko.

Platonov GV (1954): Nekotorye filosofskie voprosy diskusii o vide i vidoobrazovanii [Several philosophical issues of the discussion on species and speciation]. Voprosy Filosofii 6:116–32.

Platonov GV (1959): "Darvin, darvinizm, i filosofia" [Darwin, Darwinism, and philosophy]. Moscow: Gosudarstvennoe izdatel'stvo politcheskoi literatury.

Poliakov IM (1941): "Kurs darvinizma" [Course in Darwinism]. Moscow: Uchpedgiz.
Polianskii VI, Polianskii II (eds) (1966): "Istoriia evoliutsionnykh uchenii v biologii" [History of evolutionism in biology]. Moscow and Leningrad: Nauka.
Postanovlenie Prezidiuma Akademii nauk SSSR ot 26 avgusta 1948 goda. Po voprosu o sostoianii i zadachakh biologicheskoi nauki v institutakh i uchrezhdeniiakh Akademii nauk SSR [Decree of the presidium of the AN SSSR of 26 August 1948. On the problem of the condition and tasks of biological science in institutes and institutions of the AN SSSR]. Vestnik Akademii Nauk SSSR 9:21–24.
Prezent II (1948): Teoreticheskii put' osnovopolozhnika tvorcheskogo darvinizma [The theoretical path of the founder of creative Darwinism]. In "Sochineniia," vol 1:xx–lxiv. See Michurin 1948.
Richards RJ (1987): "Darwin and the Emergence of Evolutionary Theories of Mind and Behavior." Chicago: Univ Chicago Press.
Rukhkian AA (1953). Ob opisannom S. K. Karapetianom sluchae porozhdeniia leshchinyi grabom" [On the case of a hazel resulting from a hornbeam described by S. K. Karapetian]. Botanicheskii Zhurnal 6:885–92.
Schmalhausen II. (See Shmal'gauzen).
Semenov NN (1959): O sootnoshenii khimii i biologii [On the interrelationship of chemistry and biology]. Voprosy Filosofii 10:95–102.
Semenov NN (1961): Discussion on reorganizing the work of the USSR Academy of Sciences. In "Recent Advances in Soviet Science," London: Todd Ref Books, pp. 206–17.
Semenov NN (1962): Gumanizm nauki [The humanism of science]. In: "Nauka i Chelovechestvo" [Science and humanity], Moscow: Znanie, pp 28–38.
Semenov NN (1965): Nauka ne terpit sub'ektivizma [Science does not endure subjectivism]. NiZh 4:38–43, 132.
Semenov, NN (1966): (Emanuel NN, comp.) Akademiia nauk SSSR. Materialy k biobibliografii uchenykh SSSR. Seriia khimicheskikh nauk, no. 38. 2nd ed. Moscow: Nauka.
Semenov NN (1973): "Nauka i obshchestvo: stat'i i rechi [Science and society: articles and speeches]. Moscow: Nauka.
Serebrovskii AA (1925): Khromozomy i mekhanizm evoliutsii [Chromosomes and the mechanism of evolution]. Zhurnal eksperimental'noi biologii, Ser. B (1):49–75.
Serebrovskii AA (1928): Genogeografiia i genofond sel'skokhoziaistvennykh zhivotnykh SSSR [Genogeography and the gene fund of agricultural animals in the USSR]. Nauchnoe slovo 9:3–22.
Serebrovskii AA (1973): "Nekotorye Problemy Organicheskoi Evoliutsii" [Problems of organic evolution]. Moscow: Nauka.
Serebrovskii AA (1976): "Izbrannye trudy po genetike i selektsii kur" [Selected works on the genetics and selection of poultry]. Moscow: Nauka.
Serebrovskii AA. Problemy i method genogeografii [The method and problems of genogeography]. In "Trudy s'ezda po genetike, selektsii, i plemennomu zhivotnovodstvu" [Proceedings of the congress on genetics, selection, and animal breeding], part 2, 71–86.
Severtsov [Sewertzoff] AN (1931): "Morphologische Gesetzmässigkeiten der Evolution." Vienna: Fischer.
Shmal'gauzen II (1938): Organizm kak tseloe v individual'nom i istoricheskom razvitie [The organism as a whole in individual and historical development]. Moscow and Leningrad: Izdatel'stvo Akademii nauk SSSR.
Shmal'gauzen II (1940): "Puti i zakonomernosti evoliutsionnogo protessa" [Trends and patterns of the evolutionary process). Moscow and Leningrad: Izdatel'stvo Akademii nauk SSSR.

Shmal'gauzen II (1946a): "Faktory evoliutsii (teoriia stabiliziruiushchego otbora)" [Factors of evolution: The theory of stabilizing selection]. Moscow and Leningrad: Izdatel'stvo Akademii nauk SSSR.
Shmal'gauzen II (1946b): Problemy darvinizma [Problems of Darwinism]. Moscow: Sovetskaia nauka. [ed 2, 1969, Leningrad: Nauka.]
Shmal'gauzen II (1947a): Novoe v sovremennom darvinizme [What is new in modern Darwinism]. Priroda 12:31–44.
Shmal'gauzen II (1947b): Predstavleniia o tselom v sovremennoi biologii [Holistic ideas in current biology]. Voprosy Filosofii 2:177–83.
Shmal'gauzen II (1947c): Vneshnie faktory, mezhvidovaia bor'ba i vnutrividovoe sorevnovanie v ikh vzaimodeistii [External factors, interspecific struggle and intraspecific competition in their interaction]. In: "Vnutrividovaia bor'ba," pp 3–46.
Shmal'gauzen II (1949): "Factors of Evolution," Dobzhansky T (ed). Philadelphia: Blakiston.
Shmal'gauzen II (1964): "Proiskhozhdenie nazemnykh pozvonochnykh [The origin of terrestrial vertebrates]. Moscow: Nauka.
Shmal'gauzen II (1968a): "Kiberneticheskie voprosy biologii" [Cybernetic questions of biology], Berg RL, Liapunov AA (eds)., Kibernetika v monografiiakh, no. 4. Novosibirsk: Nauka, Sibirskoe otdelenie.
Shmal'gauzen II (1968b): "The origin of Terrestrial Vertebrates." New York: Academic Press.
Shmal'gauzen II, Formozov A, Sabinin D, Iudintsev S (1947): Nashi vozrazheniia akademiku T. D. Lysenko [Our rejoinder to academician T. D. Lysenko]. Literaturnaia gazeta, 29 Nov.
Simpson GG (1953): "The Meaning of Evolution." ed 2. New York: Mentor Books.
"The Situation in Biological Science: Proceedings of the Lenin Academy of Agricultural Sciences of the USSR, July 31–August 7, 1948" (1949): New York: Internatl Publ; Moscow: For Lang Publ House.
Strong AL (1936): The miracle maker of Odessa. Moscow News, 7 Nov.
Sukachev VN [Sukatschew W] (1928a). Einige experimentelle Untersuchungen über den Kampf ums Dasein zwischen Biotypen derselben Art. Zeitschrift für induktive Abstammungs- und Vererbungslehre 47:54–74.
Sukachev VN (1928b): "Rastitel'nye soobshchestva (vvedenie v fitosotsiologiiu)" [Plant societies (Introduction to phytosociology)]. ed 4. Leningrad and Moscow: "Kniga."
Sukachev VN (1946): Problema bor'by za sushchestvovanie v biotsenologii [The problem of the struggle for existence in biocoenology]. Vestnik Leningradskogo universiteta 2:27–39.
Sukachev VN (1953): O vnutrividovykh i mezhvidovykh vzaimootnosheniiakh sredi rastenii [On intraspecific and interspecific relations between plants]. Botanicheskii Zhurnal 1:57–96.
Sukachev VN (1954): Nekotorye itogi diskussii po probleme vida i vidoobrazovaniia i ee dal'neishie zadachi [Results of the discussion on the problem of species and speciation and its further tasks]. Botanicheskii Zhurnal 2:202–23.
Sukachev VN (1954, 1955): Obzor statei i pisem, poluchennykh Redaktsii "Botanicheskogo Zhurnal" v sviazi s diskussiei po probleme vida i vidoobrazovaniia [Survey of articles and letters received by the editors of Botanical Journal in connection with the discussion on the problem of species and speciation]. Botanicheskii Zhurnal 1:76–89; 2:217–26.
Sukachev VN (1975): "Izbrannye trudy" [Selected works]. Vol 3, "Problemy fitotsenologii" [Problems of phytocoenology]. Leningrad: Nauka.
Timoféeff-Ressovsky NW (1938): Genetica di popolazioni. La Ricerca Scientifica 11–12:7, 17–18.
Timoféeff-Ressovsky NW (1939): Genetik und Evolution. Zeitschrift für induktive Abstammungs- und Vererbungslehre 76 (1–2):158–218.

Timoféeff-Ressovsky NW, Iablokov AV, Glotov NV (1973): "Ocherk ucheniia o populiatsii" [Survey of population studies]. Moscow: Nauka.
Timoféeff-Ressovsky NW, Timoféeff-Ressovsky HA (1927): Genetische Analyse einer freilebenden Drosophila melanogaster population. Roux Archiv Entwicklungsmechanik 109(1):70–109.
Timoféeff-Ressovsky NW, Vorontsov NN, Iablokov AV (1969): "Kratkii ocherk teorii evoliutsii" [Sketch of evolutionary theory]. Moscow: Nauka.
Todes DP (1989): "Darwin Without Malthus: The Struggle for Existence in Russian Evolutionary Thought." New York: Oxford.
Toulmin S (1972): "Human Understanding." Oxford: Oxford Univ Press.
Turbin NV (1952): Darvinizm i novoe uchenie o vide [Darwinism and the new species doctrine]. Botanicheskii Zhurnal 6:798–818.
Vavilov NI (1922): The law of homologous series in variation. J Genetics 12(1):47–89.
Vavilov NI (1951): The origin, variation, immunity and breeding of cultivated plants." Chronica Botanica 13(1–6).
Vavilov NI (1959–65): "Izbrannye trudy" [Selected works]. 5 vols. Moscow and Leningrad: Izdatel'stvo Akademii nauk SSSR.
Vavilov NI (1966): "Izbrannye sochineniia: genetika i selektsia" [Selected works: genetics and selection]. Moscow.
Vavilov NI (1967). "Izbrannye proizvedeniia" [Selected works], Bakhteev FK (ed). 2 vols. Leningrad.
"Vnutrividovaia bor'ba u zhivotnykh i rastenii" [Intraspecific competition in animals and plants] (1947). Moscow: Izdanie MGU.
Vorontsov NN (1980): Sinteticheskaia teoriia evoliutsii: ee istochniki, osnovnye postulaty i nereshennye problemy [Synthetic theory of evolution: its origins, fundamental principles, and unsolved problems]. Zhurnal Vsesoiuznogo khimicheskogo obshchestva im. D. I. Mendeleeva 25(3):295–314.
Vorontsov NN (ed) (1968–1975): "Problemy Evoliutsii" [Problems of Evolution]. 4 vols. Novosibirsk: Nauka, Sibirskoe otdelenie.
Vucinich A (1970): "Science in Russian Culture, 1861–1917." Stanford: Stanford Univ Press.
Vucinich A (1984): "Empire of Knowledge: The Academy of Sciences of the USSR (1917–1970)." Berkeley: Univ Calif Press.
Zhdanov IA (1948): Tovarishchu I. V. Stalinu [To comrade I. V. Stalin]. Pravda, 7 Aug.
Zirkle C (1949): "Death of a Science in Russia: The Fate of Genetics as Described in Pravda and Elsewhere." Philadelphia: Univ Penn Press.
Zirkle C (1959): "Evolution, Marxian Biology, and the Social Scene." Philadelphia: Univ Penn Press.

New Perspectives on Evolution, pages 65-75

Current State of Evolutionary Theory in the USSR

NIKOLAI N. VORONTSOV
N K Kol'tsov Institute of Developmental Biology, USSR Academy of Sciences, Moscow, 117334 USSR

In 1935 the outstanding Soviet mathematician Andrei Kolmogorov (1903–1988) published a theoretical paper about the tempo of evolution in different kinds of populations. He demonstrated, in abstract mathematical form, that the tempo of evolution is very slow in fully panmictic populations, as you know, and also in absolutely isolated populations. The fastest tempo of evolution is in semi-isolated populations. Kolmogorov demonstrated, too, that for the tempo of evolution in such semi-isolated populations, it is not so important whether there has been a small continuous gene flow, or an alternation between absolute isolation and heavy gene exchange between such populations, or a gene flow that is periodically interrupted.

Kolmogorov's idea is a good introduction to the history of science in my country, because we have had some periods of full isolation, some periods of semi-isolation.

Soviet biology has an old evolutionary tradition rooted in Darwinism, and many Russian scientists of the older generation were active after the Revolution. Of particular prominence was Moscow University professor Mikhail Menzbir (1863–1935), who trained three very important Russian evolutionists: the great morphologist A. N. Severtsov (1863–1936); N. K. Kol'tsov (1872–1940), who introduced experimental biology into Russia; and also the paleontologist, anthropologist, and outstanding biogeographer P. P. Sushkin (1868–1928). If you study the pedigree of the Menzbir school, you will see many outstanding evolutionary biologists, geneticists, and molecular biologists in different branches of this Menzbir tree: The evolutionary embryologist Ivan Schmalhausen (1884–1963) was a student of Severtsov; Sergei Chetverikov (1880–1959), founder of population genetics, was a student of Kol'tsov; Nikolai V. Timofèeff-Ressovsky (1900–1981) was a student of Sushkin, Kol'tsov, and Chetverikov; Alexander Serebrovsky (1892–1948), the population geneticist who developed the idea of the gene pool in 1926, was a student of Kol'tsov;

population geneticist N. P. Dubinin (1907–) was a student of Serebrovsky; and so forth.

There was also a strong evolutionary tradition in the old Russian capital, St. Petersburg (Petrograd/Leningrad), especially in traditional fields such as paleontology, systematics, and botany. The tradition of the classic naturalist P. P. Semenov-Tian-Shansky (1827–1914) was carried on by his son, A. P. Semenov-Tian-Shansky, who became an entomologist. The careers of many important scientists spanned the Revolution—for example, the botanist V. N. Sukachev (1880–1967), who pioneered the study of plant communities; and ichthyologist and biogeographer Leo Berg (1876–1950), author of a theory of evolution that he called *nomogenesis* (1922).

During and after the First World War, Russia was absolutely isolated for seven or eight years. Contacts with Western scientists resumed only after our Civil War had ended, in the period 1921–1924. In those years Soviet scientists made their first visits to the West. At about the same time, in 1922, the American geneticist Hermann J. Muller made his first visit to Russia. His trip was crucial for Russian science because it stimulated Chetverikov (whose earlier work was mostly in entomology and hydrobiology) to take up drosophila genetics. Before Muller's visit, Chetverikov's group studied only the water larvae of insects; afterwards, they pioneered the study of the genetics of natural fly populations.

Something very interesting happened in the early 1920s. After the Revolution came the first wave of emigration. This wave consisted mostly of noblemen, rich people, soldiers from the White Army, artists, painters, musicians—but practically no scientists. The Crimea was the last stronghold of the White Army, and concentrated there were many brilliant Russian scientists—V. I. Vernadsky, Alexander Gurvich, and many, many others. They had the chance to evacuate with the White Army, but almost none of them did. I don't want to spread Communist propaganda, but it is a fact that at the beginning of the 1920s, although my country was hungry and certainly not rich, scientists in many different fields received support. Many palaces and estates were confiscated during this period, and it was very easy for a scientist to get a palace for scientific work. For instance, both Vavilov and Filipchenko were given palaces outside of Leningrad for their biological work, Kol'tsov got a palace in Moscow, and so forth.

We had brilliant schools of evolutionary biology in the 1920s. There were not only classical schools, such as Severtsov's school of evolutionary morphology in Moscow and Schmalhausen's group in Kiev. There were also four or five brilliant genetical schools. Nikolai Vavilov (1887–1943) headed a large group in Leningrad that included the world-renowned cytogeneticists G. D. Karpechenko (1899–1942) and G. A. Levitsky (1872–1942). The group collected cultivated plants from around the world and studied their cytogenetics, centers of origin, and hybridization. Also in Leningrad was the school of Iurii

Filipchenko (1882–1930), which included Theodosius Dobzhansky (1900–1975). In 1915 Filipchenko opened a university course on genetics and helped introduce the field in Russia. He was also interested in evolution; he wrote a book on history of evolutionary theory, and in 1927 he introduced the terms *microevolution* and *macroevolution.* His school studied the genetics of many kinds of organisms, including humans, cattle, and wheat. In Moscow, the school of Nikolai Kol'tsov focused on cell studies, cytogenetics, population genetics, and freshwater ecology. The school of Sergei Chetverikov, a branch of Kol'tsov's, studied population genetics and mutations in the field. Finally, the school of Alexander Serebrovsky—also a branch of Kol'tsov's—studied pheno- and genogeography, biostatistics, and gene structure.

Practically speaking, then, the Revolution did not interrupt the old traditions of Russian science. Before the Revolution, each Russian Ph.D. student received a stipend for a year's study in Europe. This tradition resumed in the 1920s, and many young Soviet biologists went abroad to study: in Germany were not only Timoféeff-Ressovsky and Tsarapkin, from the Chetverikov group, but also Slepkov and many others; Dobzhansky got a Rockefeller grant to study with Morgan at Cal Tech, G. D. Karpechenko spent two years working with Babcock at Berkeley; Mikhail Navashin (1896–1966) and Anton Zhebrak (1901–1965) worked in the States; V. V. Alpatov studied ecology with Raymond Pearl in Baltimore; and so on. Many Russians spent long periods working abroad until the beginning of the 1930s.

The first repression started in 1929, the Year of the Great Break. This repression was not directed against genetics, not against evolutionary biology: this was repression against the cultural strata of society. In this period, Chetverikov was arrested, Filipchenko was criticized, Kol'tsov lost his position at Moscow University but preserved his Institute of Experimental Biology. Then, at the beginning of the 1930s, some of these laboratories and departments were restored. It became almost impossible for Soviet scientists to travel abroad; practically speaking, only Vavilov could. But there was still contact. For example, Alpatov's student G. F. Gause (1910–1987) could not study with Pearl in the States, but his classic book in ecology, *The Struggle for Existence,* was published in Baltimore in 1934. Also, Western scientists were invited to come to the USSR to work. Hermann Muller worked in Russia for four or five years. Harland visited and Bridges visited, so our scientists had real contact with Western science.

Lysenkoism started in 1934–1935. It began as a fairly open discussion between scientists and uneducated Lysenkoists. Many biologists thought then that Lysenko could not be very dangerous because he was not educated. But then, in 1935, the official discussion began. It was also rather open, but soon afterwards a large group of Soviet geneticists were arrested (I. Agol, M. Levin, S. Levit, and many others), and two successive presidents of the Lenin Academy of Agricultural Sciences were arrested (Meister and Muralov)—and their place was taken by Lysenko. Then things got worse and worse. In 1940 Vavilov was arrested. He was not arrested alone: the same year, three of the four professors in the Department of Genetics at Leningrad University were arrested—Karpechenko, Levitsky, and Govorov. In the late 1930s the Lysenko mafia destroyed many people.

Paradoxically, even during the time of the Great Terror, a group of outstanding young geneticists and evolutionists survived, such as S. M. Gershenson (1906–), N. P. Dubinin (1907–), V. S. Kirpichnikov (1907), Iu. Kerkis (1907–1978), Iosef Rapoport (1912–), and many others. These were the first geneticists and evolutionary biologists of the Soviet generation. They had received normal training and had studied with well-educated and outstanding teachers in the 1920s and the early 1930s. Actually, the beginning of the Second World War prolonged the survival of genetics: people had more important things on their minds than the campaign against genetics or evolutionary biology.

During and after the war, we had good contacts with the United States and England. As for evolutionary biology, Julian Huxley came to visit Moscow and Leningrad in 1945. A small exchange of books was resumed. Some Soviet books reached the West (for example, Schmalhausen's *Factors of Evolution*), and some Western books reached my country. Just after the war, three very important books were translated into Russian. Ernst Mayr's book of 1942, *Systematics and the Origin of Species,* was translated in 1947; Erwin Schrödinger's book, *What is Life?,* was translated in 1947; and G. G. Simpson's book of 1944, *Tempo and Mode in Evolution,* was translated in 1948. The books of Julian Huxley and Theodosius Dobzhansky were about to be published, and had already reached galley proofs.

Then suddenly, in August 1948, it all came to an end. At the infamous session of the Academy of Agricultural Sciences, Lysenko triumphantly declared that he had Stalin's endorsement. There followed the full liquidation of genetics and evolutionary biology in my country—it was absolutely finished. Genetics, evolution, and general biology were routed. Laboratory vials and drosophila bottles were smashed. Laboratories, departments, and whole institutes were closed. Journals, books, and textbooks were removed from libraries and some of them were burned.

This was a black night for our science. I remember 1948 very well. At the time I was working at the Moscow Zoo, and my oldest friends were students at Moscow University. When I returned to Moscow one evening, they told me that Schmalhausen and Zavadovsky were dismissed; the next day, Formozov and Gause; each day there were more. That fall, in all universities, in all institutions, three thousand biologists lost their jobs and all possibility of research—three *thousand.*

But even then Russian biology was not finished. Genetics and evolutionary biology were absolutely destroyed—but not botany, zoology, and paleontology, and people found refuge in these classical fields of biology. It is very interesting that the first critical discussions against Lysenko started not in Khrushchev's time but in the last months of Stalin, in December of 1952. So in two classical journals edited by the botanist Sukachev, the *Botanical Journal* and the *Bulletin of the Moscow Society of Naturalists,* papers were published against Lysenko, and there was an active campaign of criticism in 1954 and 1955, during the early Khrushchev years. At this time many scientists wrote letters about the dangerous situation in our science. Three hundred biologists subscribed to one letter. Another letter was signed by outstanding Soviet nuclear physicists, including Andrei Sakharov. Yet another letter was signed by outstanding Soviet mathematicians, warning that not using statistics was dangerous not only for biology, but also for agriculture.

This was an interesting time. I was involved in things and have my own recollections. In 1953 and 1954 unofficial education in genetics resumed. At his Moscow flat, mathematician Alexei Liapunov (1911–1973) organized a seminar where young students studied problems of genetics and mathematics, and also the application of mathematics to biology (in 1948, along with genetics, mathematics had been officially excluded from biological teaching and work because mathematics and mathematicians supported Mendel). In 1956 Timoféeff-Ressovsky organized unofficial summer seminars in his biological station in Miassovo in the Urals. These two unofficial schools—the Liapunov seminar in Moscow and Timoféeff's seminar in the Urals—had a very great importance for the unofficial education for students. I was one of these students, and I ended up marrying Liapunov's daughter, E. A. Liapunova, who has been my scientific collaborator ever since. (Liapunov's home seminar and its participants are depicted in Vladimir Dudintsev's new novel, *White Robes* (Moscow 1988), but he sets it back in earlier, Stalinist times.)

In 1957 a new genetics institute—the Institute of Cytology and Genetics—was organized near Novosibirsk in Akademgorodok, the academy's new Science City. During this same period, many physicists, chemists, and mathematicians used their own protective coloration to organize cryptic support for genetics. Such cryptic laboratories were organized by the father of Soviet atomic bomb, I. V. Kurchatov. He organized a large, semi-secret genetical department at his Institute of Atomic Energy—not secret from the West, secret from the Lysenkoists. What happened in this semi-secret branch was not too easy for them to learn. The same kind of secret biological department was organized in the Institute of Chemical Physics by Nobel laureate N. N. Semenov, and so forth. In the mid-1950s, then, the situation for genetics was not too good, but it was also not too bad.

But Khrushchev came to Lysenko's support in 1958 and 1959: the new possibilities ended, the situation became dangerous again, and the last period of Lysenkoism began. In the cryptic laboratories and in the Novosibirsk institute, a little genetics survived. During this time, however, genetics was so poorly supported that there was no possibility to develop evolutionary genetics or population genetics—classical genetics, radiation genetics, but not population genetics. As for myself, in 1963 I organized a department of evolutionary biology at the Pirogov Medical Institute in Moscow and collected a group of young evolutionary biologists there (E. N. Panov, A. D. Bazykin, and N. N. Jordansky). I also organized an evolutionary seminar at the Moscow Society of Naturalists. The seminar was very active 1963–1965. Its weekly meetings and discussions not only brought evolutionists of the older and younger generations together, but it also served an important educational function, because at that time all the positions in the Department of Darwinism at Moscow University were held by Lysenkoists (F. A. Dvoriankin and V. A. Alekseev).

In October of 1964, Khrushchev was ousted in what we call the "little October Revolution" (th big one was in 1917). Now, I don't want to idealize Brezhnev or his colleagues, but during the two years following this little October revolution, we geneticists and evolutionists and biologists were given the chance to publish articles against Lysenko in scientific journals and newspapers. (My own article from this time, published in the newspaper *Komsomol'skaia Pravda,* was reprinted in the *New York Times*—this was twenty-five years ago.)

Soviet geneticists could work again after 1965 without many problems, but what had been destroyed was the educational tradition, and there was an enormous distance, a fantastic gap, between generations. Perhaps you know the population geneticist Raissa Berg (1913–). She is perhaps the youngest of the geneticists who got a normal classical education. Along with other people of my generation, I was self-educated; we never did get a normal biological education. In general, political pressure was rejected in the Brezhnev period, and genetics enjoyed more or less normal conditions for development. But evolutionary genetics was not very high on the list of priorities and was not so well supported; we did not get financial support; we did not get new departments, new laboratories, and so forth.

Nonetheless, from the mid 1960s through the mid 1980s, Soviet evolutionary biology underwent very intensive development. We now have people, coming out of the biochemical school of A. N. Belozersky, who have started to study molecular evolution (A. S. Antonov, B. S. Mednikov). We have people coming out of the protozoological schools of V. A. Dogel (1882–1955) and of his student Iurii Poliansky (1904–) who study the evolution of cells and protists (such people as I. B. Raikov, A. Iudin, and so on). We have a small group of outstanding ethologists, led by Evgenii Panov, my former Ph.D. student. We also have a large group, and a long tradition, of mathematical population and evolutionary biology. From our perspective, in studying populations it is very important to understand Markovian processes (Markov chains), which were described by the Russian mathematician A. A. Markov in 1907. His idea was developed by Kolmogorov and later by Liapunov. Some members of this biomathematical school now work in the United States, but most work in my country.

We also have many people who study comparative karyology, mostly of mammals and vertebrates, in connection with problems of chromosome speciation and isolating mechanisms. This is a particular interest of mine. There was no possibility for me to do this kind of work from 1956 to 1960, but I managed to send many species of living mammals to a colleague in Switzerland. Then in 1963–64 I organized an evolutionary laboratory at the Institute of Cytology and Genetics in Novosibirsk. In 1971 part of this laboratory moved to the Institute of Biology in Vladivostok, and the next year we organized a joint Soviet-American program, with R. Hoffman from Kansas University and C. Nadler of Northwestern University, to study speciation in some groups of Holarctic mammals using identical taxonomical, karyological, and electrophoretical methods. From 1972 to 1987 we published more than twenty-five joint papers on the evolution of ground squirrels, voles, wild sheep, and so forth. This was a unique example of Soviet-American cooperation. In contrast to other programs, it was not interrupted during any period of the cold war.

In 1985 the Gorbachev era began. I think we are seeing the development of a really progressive process in my country. We are very enthusiastic, and we hope that this process cannot be stopped. A small gene flow has begun between the USSR and the West and I am part of it—this is my first visit to the United States, at the age of fifty-five. And now, just within the last few months, a very interesting situation has developed within the Academy of Sciences in connection with the election to the new Congress of People's Deputies.

In Khrushchev and Brezhnev times, the Academy of Sciences preserved its old code. The Academy of Sciences was always a unique organization in our society in that it had secret voting, a secret ballot, in the election of new members. In the time of Khrushchev and Brezhnev, the Academy had a nucleus of really excellent scientists, and who could reject political candidates. In the summer of 1964, for example, the Academy rejected the candidacy of Lysenko's collaborator, N. I. Nuzhdin. Since that meeting, this has become a common practice, and at almost every election some candidates are rejected. This is a positive feature of the Academy.

But on the other hand, in 1929, the Year of the Great Break, there began another process—the politicization of the Academy of Sciences. At that time, under heavy political pressure, a contingent of well-educated political figures from the earlier generation was elected, such people as Bukharin and Lunacharsky. Some important scientific academicians—Vernadsky and Pavlov, for example—opposed this at the time. By the end of the 1930s, however, when the political pressure of Stalin and the Blood Terror was very intense, the Academy of Sciences elected Lysenko, and later it elected people who supported Lysenko. As for people who criticized Lysenko, or social scientists who criticized the analogues of Lysenko in the social sciences, they had no chance whatever of being elected to the academy.

So now we have two groups in the Academy of Sciences, and they are in a state of unstable disequilibrium: We have really good scientists, and we have a large group of administrators. You may have read in American newspapers—there has been very much in Soviet newspapers—about the fact that the presidium of the Academy of Sciences rejected the names of various distinguished people—Sakharov, Sagdeev, Iablokov, Ianshin, and others—for nomination as possible members of the future congress. As a result of this rejection, some eight thousand scientists organized a protest meeting near the presidium building of the Academy of Sciences. I was one of the speakers. As a result of this meeting, the voting became more democratic and most of the academy bureaucrats were rejected. And now we await the next round of Academy elections to decide the slate for the future congress.

It is interesting that two evolutionary biologists—Iablokov, who is very active in the Green movement in my country, and myself—were nominated to this congress as representatives from the scientific societies of the USSR. You may wonder why. Maybe I am not a politician. But ever since the 1940s, when I was a schoolboy, and on into the 1950s and 1960s, when I was active in scientific work, the situation in our genetics was so dangerous and politicized (and in our evolutionary biology, and in all of biology and science—because all of science is really a unity) that it was not so easy to distinguish real scientific activity from political activity.

In this time of openness, *glasnost,* and *perestroika,* I personally am looking forward to the future with great optimism. I hope that in the near future Soviet scientists can develop open contacts with the West. I hope that Western evolutionary biologists will be able to make use of the rich and diverse ideas of the various Soviet evolutionary schools, which have preserved and developed the old traditions of classical Russian biology. And I hope that the gene flow is not interrupted again.

ACKNOWLEDGMENTS

I am grateful to the Smithsonian Institution and to Dr. R. S. Hoffman for arranging my invitation to the USA; to Professor M. Adams of the University of Pennsylvania for organizing my visit to Philadelphia; to Professor L. Warren of the Wistar Institute for organizing my participation in this symposium; and to Professor C. F. Nadler (Northwestern University), N. Nadler, Dr. D. Weiner (University of Arizona), and Sally Hoffman, who helped me in many ways.

Chicago/Tucson/Washington/Philadelphia
April 1989

REFERENCES

For those interested in the development of evolutionary biology in Russia and the USSR, I can briefly comment on what is available.

Since 1965 there has been some excellent work in my country on the development of evolutionary biology. I. I. Poliansky and his brother, algologist V. I. Poliansky, published a two-volume work on the history and recent problems of evolution (*The History of Evolutionary Studies in Biology,* Moscow and Leningrad, 1966; and *Current Problems of Evolutionary Theory,* Leningrad, 1967). The history of species concepts was the subject of work by K. M. Zavadsky. His school in Leningrad concentrated a group of well-educated specialists in the history and philosophy of evolutionary biology (Y. Gall, E. I. Kolchinsky, A. B. Georgievsky, Z. M. Rubtsova, L. N. Khakina, S. A. Orlov, and others). This group at the Leningrad Branch of the Institute of the History of Science and Technology also includes an excellent specialist in the history of botany and genetics, D. V. Lebedev, who played a very important role in the campaign against Lysenko in the 1950s and 1960s. E. Lukina has published many manuscripts and diaries of K. E. von Baer. D. A. Alexandrov has studied Y. A. Filipchenko and his school. At the Moscow branch of the same institute, E. N. Mirzoian has studied the history of comparative anatomy.

These modern specialists have excellent, broadly educated predecessors in the first generation of Soviet biologists. In the 1940s and 1950s B. E. Raikov published a four-volume monograph on the first pre-Darwinian evolutionists in Russia. I. I. Kanaev, who started in the 1920s as a geneticist and zoologist (and was also a member of a circle of Leningrad intellectuals that included M. M. Bakhtin, M. V. Iudina, and others), published special monographs on the biological works of W. Goethe and on the history of comparative anatomy; his monograph *Twins* (Leningrad, 1958) was the first genetics book published after August 1948. In the mid-1930s A. E. Gaissinovich stopped his experiments on gene structure and began to study the history of biology. He published in Russian the works of Charles Naudin and Gregor Mendel; monographs on the embryologist K.-F. Wolff; and manuscripts, letters, and biographical studies of Ilya Mechnikov. He also wrote two books on the history of genetics (*The Birth of Genetics,* Moscow, 1966; and *The Birth and Development of Genetics,* Moscow, 1988), and some of his work is available in English (for example, "The Origins of Soviet Genetics and the Struggle with Lamarckism, 1922–1929," *Journal of History of Biology* 13, 1980).

More recently, the American scientist M. Adams has published outstanding works on the history of Soviet population genetics and the evolutionary synthesis in the USSR. He has made particular study of S. S. Chetverikov, N. K. Kol'tsov, and I. I. Schmalhausen ("The Founding of Population Genetics: Contributions of the Chetverikov School," *Journal of History of Biology* 1, no. 1, 1968; "Towards a Synthesis," ibid. 3, no. 1, 1970); "Science, Ideology, and Structure: The Kol'tsov Institute, 1900–1917," in L. Lubrano and S. Solomon, eds., *The Social Structure of Soviet Science,* Boulder, Colorado, 1980; "Sergei Chetverikov, the Kol'tsov Institute, and the Evolutionary Synthesis," and "Severtsov and Schmalhausen," in Ernst Mayr and William Provine, eds., *The Evolutionary Synthesis,* Cambridge, Massachusetts, 1980). Moscow geneticist V. Babkov has published a monograph on the Moscow genetical school (*The Moscow School of Evolutionary Genetics,* Moscow, 1986). D. Weiner has published a very interesting book on the traditions of Soviet schools studying the evolution of ecosystems (*Models of Nature,* Indiana, 1988).

The most important publication on the history of evolutionary biology in my country is *The Development of Evolutionary Theory in the USSR* (1917–1970s), a large book of more than 600 pages, edited by S. R. Mikulinsky and I. I. Poliansky, which came out in Leningrad in 1983. Its 70-page bibliography lists very many Soviet primary sources. This book is a good introduction to the study of Soviet evolutionary thinking. But it is necessary to note that this very important book was published before the Gorbachev era and does not include enough materials on the sharpest and most tragic chapters in the history of Soviet biology.

EDITOR'S FOOTNOTE

The following excerpts are taken from *The Situation in Biological Science* (1949, International Publishers, New York), the complete stenographic report of the Session of the Lenin Academy of Agricultural Sciences of the USSR (July 31–August 7, 1948). They include full reports by T. D. Lysenko and sixty other scientists.

Opening address by T. D. Lysenko

The party, the Government and J. V. Stalin personally, have taken an unflagging interest in the further development of the Michurin teaching. There is no more honorable task for us Soviet biologists than creatively to develop Michurin's teachings . . . (p. 49).

Concluding remarks by T. D. Lysenko

Comrades, before I pass to my concluding remarks, I consider it my duty to make the following statement. The question is asked in one of the notes handed to me, What is the attitude of the Central Committee of the Party to my report? I answer: The Central Committee of the Party examined my report and approved it (Stormy applause. Ovation. All rise) (p. 605).

Long live the Michurin teaching, which shows how to transform living nature for the benefit of the Soviet people! (applause).

Long live the Party of Lenin and Stalin which discovered Michurin for the world (applause) and created all the conditions for the progress of advanced materialist biology in our country. (applause).

Glory to the great friend of science, our leader and teacher, Comrade Stalin! (all rise, prolonged applause) (p. 617).

Draft of a letter from the session of the Lenin Academy of Agricultural Science of the USSR to J. V. Stalin

To this we are enthused by your words about progressive science, science which serves the people, science which values traditions but does not fear to raise its hand against all that is obsolete.

Hail the progressive Michurinian biological science!

Glory to the great Stalin, the leader of the people and coryphaeus of progressive science! (Stormy, prolonged and mounting applause and cheers. All rise) (p. 627).

COMMENT BY ERNST MAYR

What a pleasure it is to have a Russian evolutionary biologist among us! I hope that we will see many more, and I hope that there will also be the opposite stream. In addition to such an experience with mature scientists, I hope that we also have young Russian scientists coming over to this country. They all are most anxious to come, and it's just a question of making it possible for them. I have a personal wish, which, I'm sure, Dr. Vorontsov shares with me. Although there may be certain political difficulties, I'm one of those few people who do not know Russian, and so I have great difficulty in reading all these Russian papers. I would like that Russian scientists be permitted to publish their papers in English, as is now so customary in Germany. Most amazingly, this habit has even spread into France, where French scientists who formerly weren't permitted to say anything except in French now also publish in English. And I do hope that we can benefit from the Russians so that we can give them credit for all their achievements and new ideas. I hope that more and more of their publications in the future will be published in English.

COMMENT BY MARK B. ADAMS

The "fast tempo of evolution" described in the paper has continued. Many of the scientists mentioned were elected and are active in reform. Iablokov has become one of the USSR's most publicly prominent environmentalists. Some weeks after delivering this paper, Vorontsov was appointed to Gorbachev's cabinet as the Soviet Union's minister of the environment. He is the first non-Party figure in Soviet history to achieve ministerial rank. However, Vorontsov is a scientist first and foremost, not a politician. There has been a relatively free development of evolutionary biology since this time, and the political aspects are at present far less important than they may seem to an outside observer.

Vorontsov was also the most outstanding student of Timofèeff-Ressovsky, the brilliant Russian evolutionary geneticist. Timofèeff-Ressovsky brought the evolutionary synthesis to the European continent quite independently of what Dobzhanzky did in America.

New Perspectives on Evolution, pages 77-85

Environmental Context of Evolutionary Change: An Example From the End of the Proterozoic Eon

ANDREW H. KNOLL
Botanical Museum, Harvard University, Cambridge, Massachusetts 02138

INTRODUCTION

Natural selection, the cornerstone of the Neodarwinian paradigm, concerns the interaction between heritable variation and the environment. Population genetics provides a calculus for the description of this interaction, but it does not address either the origin of genetic novelty or the historical role of environmental variation in evolution. Over the past two decades, molecular biology has supplied an increasingly sophisticated picture of genomic change. Now, studies of the geological record promise to provide a comparably rigorous understanding of environmental change and its evolutionary consequences.

The paleontological record documents evolutionary pattern; only by inference can it be used to address process. As more than a decade of debate over punctuated equilibria has shown, multiple population genetic and/or developmental models can be advanced to explain patterns of morphological change within lineages, and perhaps only in exceptional circumstances will the record itself be sufficiently detailed to permit a choice among the alternatives (cf. Williamson 1987, Foote and Cowie 1988). But the geological record provides another type of information central to understanding how the present diversity of life came to be; it provides a perspective on the environmental context in which phenotypic innovations have arisen, spread, and/or disappeared. If the quest to understand evolution can be analogized to learning how to drive from Boston to Los Angeles, then molecular biology and population biology show us how automobiles work, while geology and paleontology indicate the necessity of turning left in St. Louis.

Regardless of organisms' (or genomes') capacity for producing phenotypic novelty, the geological record indicates quite strongly that as often as not, taxa radiate or become extinct because of changes in their environmental circumstances. The geological record of the past 550 million years (the Phanerozoic Eon, literally "age of visible animals") indicates that environmental changes of varying magnitude have been a continuing feature of the Earth's surface. Most changes in physical environment have been non-directional, with sea levels, climate, geography, and oceanographic circulation oscillating through time. The major evolutionary consequences of these oscillations have probably been the diversification of taxa as environmental opportunities expanded, followed by the removal of evolved diversity as organisms adapted for particular conditions disappeared along with their environments (e.g., Knoll and Niklas 1987 on plants; Lipps 1986 on marine microplankton).

In some cases, the removal of one dominant group has opened up evolutionary opportunities for another—the classic example being the radiation of large mammals in the wake of dinosaur extinctions—and extinction-induced replacement series appear to contribute significantly to the trajectory of "time's arrow" (Gould 1987) in evolutionary history. To the extent that environmental change has supplied a vectorial force in Phanerozoic evolution, it has done so principally through this type of replacement, as well as through directional changes in the *biological* components of environments. For example, a Mesozoic increase in the intensity of predation by shell crushing predators resulted in a directional shift in the morphologies and ecological distributions of skeletonized marine invertebrates (Vermeij 1977, 1987) and algae (Steneck 1983). The Cretaceous–Tertiary radiation of diatoms appears to have imposed a continuing selection pressure on radiolaria that have a structural requirement for silica (Harper and Knoll 1974), while the broadly contemporaneous evolution of open canopies in angiosperm-dominated forests created unprecedented opportunities for epiphytic ferns and other canopy dwellers (Knoll 1986). From the perspective of bacteria, the entire Phanerozoic diversification of animals and plants must be seen as an ever-expanding series of novel environments (e.g., Knoll and Bauld 1989).

Placing, for the moment, genotypic constraints into a black box, we can, thus, view the major patterns of Phanerozoic evolution as the unique historical product of oscillating environmental change (which may impart directionality through the agency of extinction) and directional change in the intensity of organism-organism interactions and in the biological diversity of environments.

The earlier history of the Earth is different. While Archean (>2500 Ma) and Proterozoic (2500-550 Ma) environments certainly oscillated much as in the Phanerozoic, physical environments also underwent directional change as the Earth developed from a planet with little continental crust and almost no atmospheric oxygen to one characterized by large stable continents bathed by oxygen-rich fluids. As the Earth developed, novel physical environments directly facilitated evolutionary radiations. In the limited space of this paper, I cannot begin to present detailed arguments about coupling of biological and physical evolution on the early Earth, but in order to provide something of the flavor of these interactions and the means by which paleontologists have begun to address them, I will discuss a single, rapidly developing case history of an evolutionarily important change in Earth surface environments—a complex series of tectonic and biogeochemical changes that accompanied and probably made possible the initial radiation of large animals near the end of the long Proterozoic Eon.

THE PALEONTOLOGICAL PROBLEM

In *The Origin of Species,* Darwin (1859:312–316) drew attention to the seemingly sudden appearance of diverse animals in stratigraphic horizons that today would be recognized as the base of the Cambrian System (the beginning of the Phanerozoic Eon). He saw in this paleontological pattern a serious threat to his theory and tried—unsatisfactorily, it would seem, even to himself—to explain the problematic distribution in terms of an imperfect record incompletely examined. During the past three decades, Upper Proterozoic and Cambrian sedimentary rocks have received much detailed study by geologists and paleontologists, and while fossil animals of slightly simpler organization have been found in strata slightly below those that formed the floor of Darwin's record, the problem has not disappeared. Indeed, the documentation of a long Archean and Proterozoic history of life and the concomitant realization that animals radiated rather late in the evolutionary day have, if anything, sharpened interest in the timing and rapidity of this radiation.

The oldest sedimentary rocks that one can examine profitably for evidence of biological activity are approximately 3500 Ma old successions found in South Africa and Australia. Microfossils, stromatolites, and geochemical signatures found in both sequences indicate that by the time these rocks were deposited, anaerobic prokaryotes had already diversified to form ecologically complex communities fueled by photosynthesis and including populations with complementary metabolisms capable of cycling biogeochemically important elements (for a recent review, see Schopf 1983). During late Archean to early Proterozoic times (2700–2200 Ma), stably oxic environments spread across the globe, and aerobic Eubacteria and, presumably, Archaeobacteria proliferated (Cloud 1974, Knoll 1979, Schopf 1983).

The time of origin of eukaryotic microorganisms is more difficult to ascertain, but both microfossils and biomarker organic geochemical evidence indicate that by 1800–1600 Ma ago, eukaryotic algae had become significant features of communities in many parts of the world (see, for example, references cited in Knoll 1990). This date constitutes a minimum time of appearance for protists containing both mitochondria and chloroplasts and does not necessarily reflect the evolutionary origin of the eukaryotic cell. Problems of preservation may hinder recognition of an earlier algal radiation; heterotrophic protists with and, still more primitive, without mitochondria (Sogin et al. 1989) must have originated earlier.

Macroscopic algae, or seaweeds, occur in 1400 Ma-old sediments from China, North America, and Australia (Walter et al. 1976, Du et al. 1986, Grey and Williams 1990), and green, red, and possibly brown algae of considerable morphological complexity are widely distributed in 700–900 Ma old rocks (e.g., Hofmann 1985; Butterfield et al. 1988; Butterfield, Knoll, and Swett, unpublished data). Thus, it cannot be argued that complex eukaryotic multicellularity *per se* originated with macroscopic animals.

Despite the impressive antiquity of prokaryotes, unicellular protists, and seaweeds, there is no unequivocal paleontological evidence for large animals in rocks older than ca. 600 Ma (Cloud 1968, Runnegar 1982a, Glaessner 1984). Once again, this is not equivalent to a statement that the kingdom Animalia originated at this time. Various molecular and geological arguments have been advanced in favor of an 800–1000 Ma origin of animal phyla (e.g., Runnegar 1982a, Walter and Heys 1985); however, even if one accepts these arguments, it is clear that the dramatic appearance of animal body and trace fossils near the top of the Proterozoic is not a stratigraphic or preservational artifact. it records a genuine radiation of *macroscopic* animals.

BIOLOGICAL SUGGESTIONS OF LATE PROTEROZOIC ENVIRONMENTAL CHANGE

Life originated 3000 Ma or more before the Ediacaran radiation of large animals; single-celled protists appeared more than 1000 Ma before this event; and even morphologically complex metaphytic algae have a long pre-Ediacaran history. In attempts to explain the late radiation of macroscopic animals, biologists have repeatedly turned to latest Proterozoic environmental change for a solution (e.g., Nursall 1959, Raff and Raff 1970, Cloud 1976, Runnegar 1982a). A popular suggestion has been that prior to 600 Ma ago, the atmosphere and hydrosphere contained too little oxygen to support the physiology of large animals. [An alternative oxygen-control hypothesis which held that prior to 600 Ma ozone levels were insufficient to protect animals from lethal radiation (Berkner and Marshall 1965) appears to be untenable; current models suggest that essentially full protection from UV-C radiation was established at least 1000 Ma earlier (e.g., Kasting 1987)]. Runnegar (1982b) for example, has estimated that *Dickinsonia,* a large, flat invertebrate found in Ediacaran assemblages, required atmospheric oxygen levels at least 6–10% of the present-day concentration in order to ensure adequate oxygenation of its tissues. Larger, thicker animals (which exist in some latest Proterozoic assemblages, Fedonkin 1985) or animals without well-developed circulatory systems would require higher $pO2$.

GEOLOGICAL INDICATIONS OF LATE PROTEROZOIC ENVIRONMENTAL CHANGE

If atmospheric and hydrospheric oxygen concentrations did indeed increase just prior to the Ediacaran radiation, what imprint would this event have left in the geological record? The question is crucial, because without geological evidence, the entire hypothesis must remain speculative. I have discussed the late Proterozoic record of environmental change in detail elsewhere (Knoll 1990); here I will confine my remarks to a brief summary of accumulating evidence that supports the hypothesis of an immediately pre-Ediacaran rise in pO2.

Because the capacity for an increase in atmospheric oxygen is generated principally by the burial of photosynthetically produced organic matter, the first approach to the problem is to ask whether or not the late Proterozoic era was characterized by anomalously high rates of organic carbon burial. Stable carbon isotopic ratios in carbonates and co-existing organic matter document rates of organic carbon burial at the time of their formation (Hayes 1983, Holser et al. 1986), and both broad isotopic surveys (Eichmann and Schidlowski 1975, Strauss 1988) and detailed, stratigraphically and paleoenvironmentally controlled investigations of 600–800 Ma-old successions (Knoll et al. 1986, Fairchild and Spiro 1987, Kaufman et al. 1990) indicate that the immediately pre-Ediacaran epoch was a time during which extended intervals of strongly enhanced organic carbon burial alternated with shorter periods of lower accumulation rates. Independent geological features suggest an explanation for this pattern. The sedimentary burial of organic matter is sensitive to basinal geometries, oceanic circulation patterns, and rates of overall sedimentation. 800 to 900 Ma ago, one or a small number of late Proterozoic supercontinents began to break apart, with the formation of what may be a uniquely large number and extent of rapidly subsiding extensional basins. Most of these basins were flooded by marine waters. Their rapid sedimentation rates and long, narrow bathymetric profiles, in combination with warm, oxygen-poor deep circulation, would have facilitated organic carbon burial. Some and perhaps all of the intervals of reduced organic carbon burial correspond stratigraphically to late Proterozoic ice ages—periods of lowered sea level and vigorous circulation of cold, oxygen-rich bottom waters. Additionally, strontium isotopic ratios in carbonate rocks suggest relatively strong hydrothermal input to oceans at this time, providing a significant source of reduced materials that would facilitate organic carbon burial on the sea floor (Veizer et al. 1983; Derry et al. 1989).

The summary points are (1) that the immediately pre-Ediacaran epoch was an interval during which high rates of organic carbon burial must have generated a great deal of oxygen, and (2) that this burial can, in turn, be related to tectonic and climatic patterns known to have characterized the late Proterozoic era. Oxidizing power generated does not necessarily equate with atmospheric oxygen accumulated because the oxygen can be removed by the oxidation of reduced sulfur and iron, as well as sedimentary organic matter. Much of the oxygen generated by late Proterozoic organic carbon burial must have been consumed in this way, and there is as yet no reliable means of quantifying the net build-up of pO2; however, various lines of evidence support the hypothesis that pO2 did increase significantly at or just prior to the initial radiation of macroscopic animals. These include the apparent decoupling of secular variation in the isotopic records of sulfur and carbon in pre-Ediacaran successions (Lambert and Donnelly 1989) (implying that additional elements, probably ferrous iron, played an important role in the redox balance of earlier Proterozoic oceans), the occurrence of sedimentary iron formation in sub-Ediacaran stratigraphic successions (Young 1976), and approximately contemporaneous changes in the biomarker molecular signatures of preserved organic matter (Summons et al. 1988). All of these lines of evidence are incompletely documented, but collectively they present tantalizing preliminary support for the idea of biologically important latest Proterozoic environmental transition.

Oxygen-rich environments undoubtedly constituted a necessary condition for the Ediacaran radiation, but they would not in and of themselves have provided sufficient conditions for this event—the evolution of large animals also required a sophisticated system of developmental control. The close temporal correlation between major environmental change and the observed radiation indicates that the necessary genetic controls were either easily evolved or, perhaps more likely, already in place in minute, nematode-grade animals able to live in earlier Proterozoic environments with limited oxygen.

DISCUSSION

The paleontological record has long indicated that the end of the Proterozoic Eon was a time of major biological innovation. Now, it is becoming clear that this epoch was equally a period of profound tectonic, climatic, and biogeochemical change. While the nature of late Proterozoic interactions between the evolving Earth and its biota are not yet completely clear, enough data have accumulated to permit the articulation of the hypothesis that a tectonic event (the incipient breakup of a late Proterozoic supercontinent) affected the fluxes of biogeochemical cycles, leading to an increase in pO2 that made possible the evolution of large animals (Knoll 1990). This hypothesis accounts for a wide variety of geological and paleontological data, and, to the best of my knowledge, does not run counter to existing facts. Nonetheless, for the present, its value lies principally in its suggestion of a research agenda rather than in any perceived Truth.

The hypothesis makes clear the need for further carbon and sulfur isotopic measurements in middle Proterozoic to Cambrian sediments and makes predictions about the patterns that should be detected. It suggests the importance of examining late Proterozoic ocean bottom sediments (now preserved only as metamorphosed slivers in ophiolite belts) and predicts a number of their features. It suggests that we need to know more about late Proterozoic stratigraphic and paleoenvironmental patterns in the sedimentary distribution of iron, as well as secular variations in geochemical indicators of oxidation, and predicts how these patterns should change near the end of the Proterozoic Eon.

It invites biogeochemical comparisons between the late Proterozoic and the latest Paleozoic-Mesozoic period when Pangaea broke apart. In short, although the hypothesis is historical in nature, it is amenable to tenting in that it makes explicit predictions about patterns observable in the geological record.

Perhaps the single most important feature of this hypothesis of coupled late Proterozoic biological and environmental change is its form, its explicit articulation of the view that biological and physical Earth evolution are linked via the one system in which they both participate—biogeochemical cycles. While the genetic and population-level processes that propel biological evolution may be independent of environmental history, large-scale historical patterns of evolution clearly are not. Earth's planetary development has both influenced and been influenced by biological evolution in profound ways. Regardless of whether or not the particular hypothesis advocated here survives its geological tests, there can be little doubt that a satisfactory understanding of the paleontological record can come only when it is interpreted in the context of ongoing environmental variation. This rejuvenated form of the old partnership between paleontology and geology promises to teach us much of importance to paleontology's other partner, evolutionary biology.

ACKNOWLEDGMENTS

Research leading to the development of the ideas expressed in this essay was supported by grants from NASA, NSF, and the John Simon Guggenheim Foundation.

REFERENCES

Berkner LV, Marshall, LC (1965): On the origin and rise of oxygen concentration in the earth's atmosphere. J Atmos Sci 22:225.

Butterfield NJ, Knoll AH, Swett K (1988): Exceptional preservation of fossils in an Upper Proterozoic shale. Nature 334:424.

Cloud P (1968): Pre-metazoan evolution and the origins of metazoa. In Drake T (ed): "Evolution and Environment." New Haven: Yale Univ Press, p 1.

Cloud P (1974): Evolution of ecosystems. Am Sci 62:54.

Cloud P (1976): Beginnings of biospheric evolution and their biogeochemical consequences. Paleobiology 2:351.

Darwin C (1859): "On the Origin of Species by Means of Natural Selection." London: J Murray. [Reprinted 1979, New York: Avenel Books.]

Derry L, Keto LS, Jacobsen S, Knoll AH, Swett K (1989): Sr isotopic variations of Upper Proterozoic carbonates from East Greenland and Svalbard. Geochim Cosmochim Acta 53:2331.

Du R, Tian L, Li H (1986): Discovery of megafossils in the Gaoyuzhuang Formation of the Changchengian System, Jixian. [In Chinese, with English summary.] Acta Geol Sinica 1986:115.

Eichmann R, Schidlowski M (1975). Fractionation between coexisting organic carbon-carbonate carbon pairs in Precambrian sediments. Geochim Cosmochim Acta 39: 585.

Fairchild IJ, Spiro B (1987): Petrological and isotopic implications of some contrasting Late Precambrian carbonates, NE Spitsbergen. Sedimentology 34:973.

Fedonkin MA (1985): Precambrian metazoans: Problems of preservation, systematics and evolution. Phil Trans Roy Soc, London 311B:27.
Foote M, Cowie RH (1988): Developmental buffering as a mechanism for stasis: Evidence from the pulmonate *Theba pisana.* Evolution 42:396.
Glaessner M (1984): "The Dawn of Animal Life: A Biohistorical Study." Cambridge: Cambridge Univ Press.
Gould SJ (1987): "Time's Arrow, Time's Cycle." Cambridge MA: Harvard Univ Press.
Grey K, Williams IR (1990): Problematic bedding-plane markings from the Middle Proterozoic Maganese Subgroup, Bangemall Basin, Western Australia. Precambri Res 46:307.
Harper HE, Knoll AH (1974): Silica, diatoms, and Cenozoic radiolarian evolution. Geology 3:175.
Hayes JM (1983): Geochemical evidence bearing on the origin of aerobiosis, a speculative hypothesis. In Schopf JW (ed): "Earth's Earliest Biosphere: Its Origin and Evolution," Princeton: Princeton Univ Press.
Hofmann HJ (1985): The mid-Proterozoic Little Dal macrobiota, Mackenzie Mountains, north-west Canada. Palaeontology 28:331.
Holser WT, Magaritz M, Wright J (1986): Chemical and isotopic variations in the world ocean during the Phanerozoic. In Walliser O (ed): "Global Bio-Events." Berlin: Springer-Verlag, p 63.
Kasting JF (1987): Theoretical constraints on oxygen and carbon dioxide concentrations in the Precambrian atmosphere. Precambr Res 34:205.
Kaufman AJ, Hayes JM, Knoll AH, Germs GJB (1990): Carbon-isotopic abundances in carbonates and organic matter from Upper Proterozoic successions in Namibia: Stratigraphic, diagenetic, and metamorphic Effects. Precambr Res.
Knoll AH (1979): Archean photoautotrophy: Some alternatives and limits. Origins Life 9:313.
Knoll, AH 1986: Patterns of change in plant communities through geological time. In Diamond J, Case TJ (eds): "Community Ecology." New York: Harper & Row, p 126.
Knoll AH (1990): Biological and biogeochemical preludes to the Ediacaran radiation. In Lipps J, Signor P (eds): "Origins and Early Evolutionary History of the Metazoa." New York: Plenum Press.
Knoll AH, and Bauld J (1989): The evolution of ecological tolerance in prokaryotes. Trans Roy Soc Edinburgh Earth Sci 80:209.
Knoll AH, Hayes JM, Kaufman AJ, Swett K, Lambert IB (1986): Secular variation in carbon isotope ratios from Upper Proterozoic successions of Svalbard and East Greenland. Nature 321:832.
Knoll AH, Niklas KJ (1987): Adaptation, plant evolution, and the fossil record. Rev Palaeobot Palynol 50:127.
Lambert IB, Donnelly TH (1989): The paleonenvironmental significance of trends in sulfur isotope composition in the Precambrian: A review. In Herbert HK (ed): "Stable Isotopes and Fluid Processes in Mineralization." Canberra: Spec Publ Geol Soc Austral.
Lipps J (1986): Extinction dynamics in pelagic systems. In Elliott DK (ed): "Dynamics of Extinction." New York: John Wiley and Sons, p 87.
Nursall, JR (1959): Oxygen as a prerequisite to the origin of the metazoa. Nature 183:1170.
Raff, RA, Raff EC (1970): Respiratory mechanisms and the metazoan fossil record. Nature 228:1003.
Runnegar B (1982a): The Cambrian explosion: Animals or fossils? J Geol Soc Australia 29:395.
Runnegar B (1982b): Oxygen requirements, biology and phylogenetic significance of the late Precambrian worm *Dickinsonia,* and the evolution of the burrowing habit. Alcheringa 6:223.

Schopf JW (ed) (1983): "Earth's Earliest Biosphere: Its Origin and Evolution." Princeton: Princeton Univ Press.
Sogin ML, Gunderson JH, Elwood HJ, Alonso RA, Peattie DA (1989): Phylogenetic meaning of the kingdom concept: An unusual ribosomal RNA from *Giardia lamblia.* Science 243:75.
Steneck RS (1983): Escalating herbivory and resulting adaptive trends in calcareous algae. Paleobiology 9:45.
Strauss H (1988): Proterozoic organic carbon—preservation and record. Abstr Symp: "The Proterozoic Biosphere: An Interdisciplinary Study." Los Angeles, p 31.
Summons RG, Powell TG, Boreham CJ (1988): Petroleum geology and geochemistry of the Middle Proterozoic McArthur Basin, northern Australia. III. Composition of extractable hydrocarbons. Geochem Cosmochim Acta 51:1747.
Veizer J, Compston W, Clauer N, Schidlowski M (1983): $^{87}Sr/^{86}Sr$ in Late Proterozoic carbonates: Evidence for a "mantle event" at 900 Ma ago. Geochim Cosmochim Acta 47:295.
Vermeij GJ (1977): The Mesozoic marine revolution: Evidence from snails, predators and grazers. Paleobiology 3:245.
Vermeij GJ (1987): "Evolution and Escalation." Princeton: Princeton Univ Press.
Walter MR, Heys GR (1985): Links between the rise of the metazoa and the decline of stromatolites. Precambr Res 29:149.
Walter MR, Oehler JH, Oehler DZ (1976): Megascopic algae 1300 million years old from the Belt Supergroup, Montana: A reinterpretation of Walcott's *Helminthoidichnites.* J. Paleontol 50:872.
Williamson PG (1987): Selection or constraint? A proposal on the mechanism for stasis. In Campbell KSW, Day MF (eds): "Rates of Evolution." London: Allen & Unwin, p 129.
Young GM (1976): Iron formation and glaciogenic rocks of the Rapitan Group, Northwest Territories. Precambr Res 3:137.

New Perspectives on Evolution, pages 87-99

The Species as a Unit of Large-Scale Evolution

STEVEN M. STANLEY
Department of Earth and Planetary Sciences, The Johns Hopkins University, Baltimore, Maryland 21218

INTRODUCTION

In 1971, Ernst Mayr stated, "Species are the real units of evolution." He went on to say, "Without speciation [evolutionary branching events] there would be no diversification of the organic world, no adaptive radiation, and very little evolutionary progress. The species, then, is the keystone of evolution." At the time that Mayr wrote these words, species were seldom being used as units in the study of macroevolution.

Macroevolution has been defined in many ways, but most people understand it to mean evolution that transcends species boundaries. Macroevolution is most commonly studied by paleontologists, who have at their disposal information about the history of biological taxa over millions of years but who cannot study populations from generation to generation. As late as the early 1970s, it was the convention in paleontology to study macroevolution using higher taxa, usually genera or families, as units of analysis. This approach was favored by George Gaylord Simpson, who wrote in his highly influential book (1953), "Species can sometimes be used to advantage in studies of taxonomic rates of extinct groups, but genera are usually the most useful units at present." The reason that paleontologists tended to employ higher taxa to study rates, trends, and patterns of evolution was that a relatively small percentage of extinct species have been preserved and recognized in the fossil record. The percentages of genera and families that have been preserved and identified are greater, and yet these higher taxa offer rather weak resolution. For example, we recognize that a genus or family has become extinct only after it has lost all of its species, so that a very large taxon of this type may suffer almost total extinction of its species and yet be tallied as a survivor. When we simply record its survival, we fail to note that a severe catastrophe befell it. Similarly, if several families enjoy great diversification in number of species over a period of time without giving rise to new families, and if we measure rates at the family level, we may fail to note any expansion whatever. Fortunately, it is now possible to improve our resolution when working at the species level. We can accomplish this by circumventing certain preservational problems that would ordinarily stymie us. One way of doing this is to look to portions of the fossil record that are of especially high quality and that reveal the history of species in unusual detail. A second approach is to employ tricks to avoid the inadequacies of the fossil record. All of us who are engaged in efforts of this type are making use of improved taxonomic data and the improved dating of fossils that has been afforded by modern stratigraphic techniques.

STABILITY OF SPECIES

Let us envision an imaginary world in which species arise almost instantaneously and then experience no further change. Under such circumstances species would make ideal units for the study of macroevolution. Microevolution, or evolution within species, would be concentrated in the origin of species, and paleontologists could comfortably focus upon the fates of species after they arose—their survival through geologic time and their generation of descendant species by way of speciation events. Of course, the world is not so simple. Species do evolve after they have formed. Nonetheless, if species typically form rapidly from their ancesters, by branching events, and then enter into long intervals of minor change before becoming extinct, they should serve as reasonably good units of large-scale evolution. The notion that species do, in fact, behave in this fashion was advanced in the early 1970s as the punctuational model of evolution (Eldredge 1971; Eldredge and Gould 1972). This model contrasts with the traditional, gradualistic model, which holds that most evolutionary change occurs by the long-term transformation of established species. Even if one adopts a gradualistic view, lineages can serve as useful units for the study of macroevolution. They are simply less than perfect units because they change their character in major ways in the course of geologic time. In this light, we can note that George Gaylord Simpson, in 1951, defined evolutionary species as being valid taxa because each has "its own unitary role and tendencies." An evolutionary species is one that changes significantly through time. More commonly we refer to such an entity as an evolutionary lineage, or a chronospecies. Lineages, whether they are approximately static or undergo substantial change, represent elements in the diversity of higher taxa. They may branch (speciate) or suffer extinction in the course of a given interval of geologic time. The study of rates of speciation and extinction is crucial to our understanding of changes in the composition of higher taxa. Some of these changes constitute trends on a macroevolutionary scale. Others represent patterns of change and diversity, and often these transform the global ecosystem.

It was my early conversion to the punctuational model of evolution that led me to develop a strong interest in using species as units of macroevolution. My conversion to the punctuational model resulted from my analyses of species durations in geologic time (Stanley 1975a, 1979). What I found from examining large volumes of published data was that the average species duration in many higher taxa of animals and plants has been in the order of 1 to 20 million years. In terms of population genetics, this translates into 100,000 to 10,000,000 generations or so. Such spans of time provide enormous opportunities for gradual change to occur, and yet in many groups the change that we typically observe over such intervals has been utterly trivial. A criticism of this point of view has come from population geneticists, who have claimed that the identity of a fossil species is in the eye of the beholder; in other words, it is a figment of some taxonomist's imagination. Certainly, taxonomy at the species level is flawed when only fossils are available for study, but this obscures the crucial issue. If a competent taxonomist, confronted with excellent fossil material, identifies populations separated by great spans of geologic time as a single species, these populations do not encompass appreciable change. Our yardstick here has to be the intervals of time required for the origins of higher taxa (genera and families)—and the adaptive transitions that such origins entail. In this light, I have been struck especially by the remarkably rapid adaptive radiations documented by the fossil records of many groups of animals and plants. Many higher taxa have diversified markedly at the genus and family level over periods of just a few million years. These same higher taxa display mean species longevities measured in millions of years. Clearly, the differentiation displayed was not primarily the product of the gradual transformation of well-established species. Under these circumstances, we have no recourse but to invoke rapidly divergent speciation to account for most of the evolutionary change that has occurred. The punctuational model simply focuses on rapid evolutionary change, which is to say, change concentrated in brief geological intervals. In 1954, Ernst Mayr suggested that our failure to document the appearance of many evolutionary novelties in the fossil record may result from the fact that most such changes occur by way of rapid speciation events that entail small populations in local areas. Even if the fossil record of well-established species is often very good, the fossil record of rapidly divergent speciation events is bound to be abysmal.

I continue to defend the use of species longevities, as documented by the fossil record, to oppose the gradualistic model of evolution. Even if fossil data that are taken to represent a nearly static condition for a species over millions of years actually represent several species that cannot be distinguished in the fossil record, a strong case remains. Biologists refer to nearly identical species as sibling species. If a static fossil lineage includes not a single species but ten or twelve sibling species, then we simply have an example of many static lineages rather than just one. The punctuational case is actually strengthened. Nonetheless, spurred by skepticism of some biologists who were uncomfortable with the subjectivity of taxonomy, I embarked a few years ago on a different approach that made use of fossil data for the group that I know best, the bivalve mollusks. The effort here was to avoid taxonomic designations and simply to collect a large unbiased sample of data on rates of morphologic change. In the literature a vast quantity of data exists purporting to represent rates of change in fossil morphology. Unfortunately, an assessment of these published rates is meaningless. First of all, many of the so-called rates have not been shown to represent true lineages. In most calculations of these rates researchers have simply assumed that a particular population descended from an older one by the gradual transformation of an entire species. This unjustified connect-the-dots approach has yielded data of dubious significance. In some cases, the transition may have been gradual, without branching, but in other cases the younger population has represented a distinct lineage that formed from the lineage of the first population by a speciation event. Another problem is that, as Eldredge and Gould 91972) have put it, workers have tended to ignore the fact that "stasis is data." In other words, paleontologists have tended to calculate and publish rates only where they thought they saw significant change in advance. The third problem is that most published rates represent nothing but differences in body size. Though sometimes having ecological significance, this kind of change is quite trivial compared to the more profound changes in morphology and adaptation associated with transitions from genus to genus or family to family.

We must avoid biases such as the ones that I have described. We must demonstrate that we are almost certainly dealing with real evolutionary lineages, not pairs of lineages that might have been separated by speciation events. We must also employ large, unbiased sets of lineages in order to assess the general pattern of evolution. Finally, we must examine large, unbiased sets of morphologic characters, preferably sets that depict morphology in a comprehensive way. Ideally, we should employ multivariate statistics.

In an effort to follow these guidelines my student Xiangning Yang and I studied 19 arbitrarily chosen lineages of marine bivalve mollusks (Stanley and Yang 1987). For each we measured 24 morphological characters, which represented a wide range of adaptive traits. Our primary goal was to determine the degree to which 4 million-year-old populations of these lineages differ from living populations. As a yardstick, we measured geographic variability within Recent populations, using the same variables. The particular lineages that we studied were assembled according to several arbitrary rules, so that the results would not be biased by preconceptions. Each of the populations representing one of the lineages was compared to one or more living populations that belonged to the extant species most closely resembling the fossil populations. In 12 of the 19 cases, it turned out that the fossil and Recent populations have traditionally been assigned the same specific name. Our multivariate morphometrics revealed a pattern of approximate evolutionary stasis. The ancient populations did not differ from Recent populations appreciably more than conspecific living populations differ from each other. Furthermore, we were able to trace three of the lineages back to 17 million-year-old populations that also turned out not to differ appreciably in form from modern populations.

While our study of marine bivalves revealed an overwhelming prevalence of evolutionary stasis, it did not permit us to identify the locus of evolutionary change that produced the lineages that we studied. Since the original formulation of the punctuational model, it has been suggested that a large percentage of evolutionary change may occur quite rapidly in geological time, but within well-established species. The result would be a kind of staircase pattern of evolution, with long intervals of approximate stasis being separated by rapid steps of substantial change. In fact, this pattern has been documented for certain taxa such as the planktonic foraminiferan genus *Globorotalia* (Malmgren et al. 1983). Even so, there have been studies similar to ours that have pointed to speciation events as the locus of most evolutionary change. For example, Cheetham (1986) conducted a detailed multivariate analysis of the marine bryozoan genus *Metrarabdotos* in the Caribbean region and found that individual species were remarkably static over millions of years and must have arisen by way of rapid divergence. This conclusion was based on the fact that many species overlapped with the species from which they evolved. Furthermore, the system was diversifying, which means that there were very few ancestral species that might have been transformed into new species by way of staircase transitions. A second study yielding the same kind of result was that of Cronin (1987), who discovered marked evolutionary stasis in the marine ostracode genus *Puriana.* This genus also diversified through time, and nearly all pairs of ancestral and descendent species overlapped in geologic time.

There have been other studies, such as that of Sheldon (1987) for Silurian trilobites, that have demonstrated some gradual morphological changes within lineages. These studies, however, have generally focused on just one or a very few morphologic traits. Had a large, random sample of traits been studied, a prevalence of evolutionary stasis might well have been observed.

When we employ species as units of analysis in macroevolution, we are in many ways treating them in the ways that population biologists treat individual organisms. Instead of birth and death, we study speciation and extinction. Thus, it is readily apparent that we can benefit from the use of demographic analogies. Of special interest are rates of speciation and extinction. How do they vary among taxa or among sets of species that differ biologically from one another? How do they vary through time within particular groups of species?

Differential rates produce trends and patterns, and these suggest the nature of the factors that control the rates themselves.

LARGE-SCALE TRENDS

When we contemplate the origin of large-scale evolutionary trends through the appearance and disappearance of certain kinds of species, analogies with population biology are readily apparent. We can envision the origin of trends in three different ways (Stanley 1979). The first mode of transformation that we can consider is analogous to natural selection within a population, in which the individual organism is the unit and differential rates of birth and survival yield evolutionary trends. At the higher level, the species is the unit and the analogous processes are speciation and extinction. Strictly speaking, the analogy here is with selection at the population level for asexual organisms, because species, by definition, do not interbreed. In the higher-level process, which I have termed species selection, a particular kind of species may be favored either because it tends to speciate at a high rate or because it tends to experience a low rate of extinction (Fig. 1). Possessing both traits represents the best of both worlds, but either trait, if pronounced enough, can lead to the success of a particular kind of species. The concept of species selection builds on the idea that species in macroevolution are comparable to mutations in microevolution (Wright 1956; Mayr 1971). As Mayr has put it, "every new species is an ecological experiment." In other words, the fact that a new species forms under particular circumstances is no guarantee that the species will fare well in a broader evolutionary context. It may suffer rapid extinction, or it may survive for a considerable time but fail to leave descendent species.

In short, speciation produces the raw material of large-scale evolution. Inasmuch as the fossil record provides only fragmentary evidence of most large segments of the tree of life (phylogenies), it is not easy to analyze species selection in detail. The easiest way to do so is to evaluate selection between two sets of species which represent alternative character states (Stanley 1981). This kind of selection is equivalent to natural selection at the individual level between the two morphs within a dimorphic population of individuals. Here, we can recognize the ascendancy of one character state without having a complete picture of phylogeny. One example is the expansion of siphonate burrowing marine bivalve mollusks and the decline of non-siphonate species (Stanley 1981, 1986). This trend has almost certainly been driven by the greater ability of siphonate species to withstand predation, through their more rapid and deeper burrowing, and we can demonstrate that predation has become more intensive in the course of geologic time. In fact, we can show that siphonate species during the past few million years have experienced much greater geologic longevities than non-siphonate species, whose populations are smaller and less stable.

One general application of species selection is in the explanation of the prevalence of sexual reproduction in nearly all groups of higher plants and animals. The puzzle of sexuality is that it entails evolutionary sacrifice on the part of females, who mix their genetic legacy with that of males rather than contributing their genotypes in toto to their offspring. Thus, asexual forms appear to have a short-term advantage in natural selection. It is also apparent that asexual taxa have arisen many times in the evolution of higher plants and animals. What is quite clear from the distribution of asexual forms in the

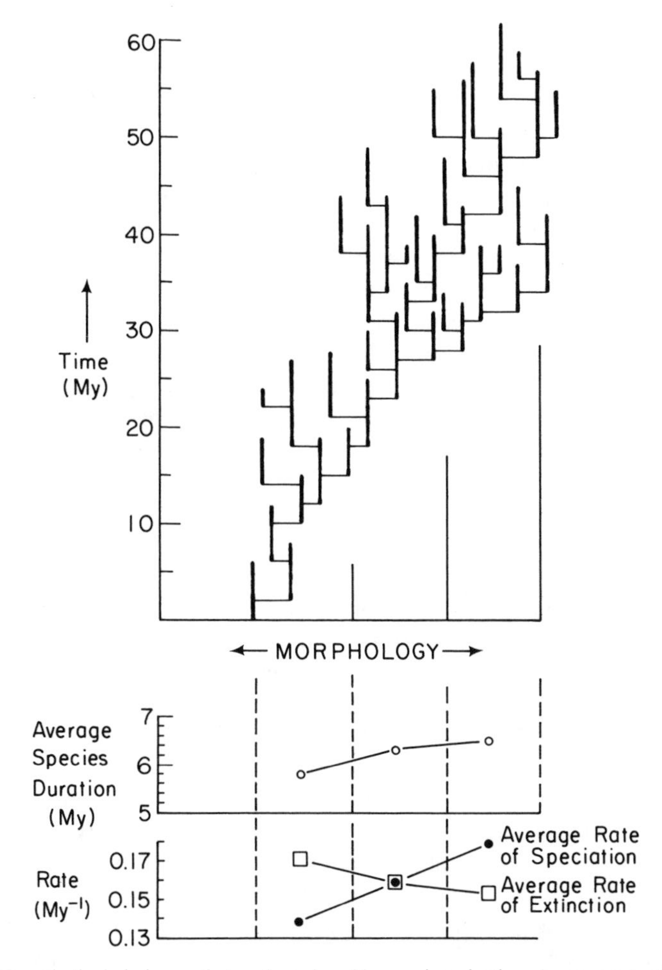

Fig. 1. Hypothetical phylogenetic trend produced by species selection. Average rate of speciation increases toward the right, and average rate of extinction, which is inversely proportional to average species in duration, decreases in the same direction. Speciation events moving to the left are equal in number to those moving to the right and contribute the same total amount of morphologic change, so that direction of speciation plays no role in the formation of the trend (from Stanley 1979). My = million years.

modern world is that they tend to be small groups that have never diversified appreciably. Cogent arguments have been made that sexual species have an advantage over asexual forms in survival against extinction and in propensity to diversify. Sexual reproduction allows species to adjust to environmental changes by experiencing high rates of recombination, which provide raw material for natural selection. Even if most species do not evolve markedly during their existence, as I believe, they may benefit from the adaptive flexibility that sexuality confers. It may allow them to make minor adjustments to changes of environmental conditions in order to ward off extinction (Maynard-Smith 1978). In addition, sexuality provides for the elimination of deleterious mutations that can accumulate in asexual clones (Muller 1964; Felsenstein 1974). Sexual reproduction also promotes diversification (Stanley 1975b). Asexual taxa do not truly speciate, they simply expand their set of adaptations slowly, through the accumulation of new mutations. Recombination clearly plays a major role in speciation, allowing small populations to move rapidly in new directions. Thus, sexual species are favored in nature both by their resistance to extinction and by their propensity to speciate. Asexual forms evolve from sexual forms only rarely, because transitions of this type are biologically difficult (Nunney 1989), and they tend not to survive long because they experience high rates of extinction and low rates of diversification in comparison to sexual species.

Just as population biologists recognize the existence of genetic drift within populations (random changes in the gene pool) we can envision the occurrence of phylogenetic drift at the level of macroevolution (Stanley 1979). Drift at both levels is difficult to recognize in nature simply because the pattern that it follows is the same as that produced by selection. Drift simply entails accidental changes rather than changes generated by adaptive disparities. Because of the statistics of small numbers, genetic drift is most common within small populations. Similarly, phylogenetic drift is to be expected most commonly within small segments of phylogeny.

The third mode of macroevolutionary change can be termed directed speciation. This process is crudely equivalent to mutation pressure within a population, which produces changes in the composition of the population independent of natural selection. Directed speciation represents a tendency for certain kinds of species to form in the course of geologic time while others are less likely to appear. Directed speciation can result from a morphogenetic tendency for certain types of evolution to occur. Thus, we commonly see certain types of morphologic features evolving many times with a taxon, while others, which would be envisioned to be highly advantageous, almost never appear. Another way in which direct speciation can develop is through a process that can be called isolate selection (Stanley 1979). Here, even if small populations evolve in many different directions from a parent population—that is, they begin to evolve into new species—some of these forms may be of types that preferentially blossom to form full fledged species, whereas others may be of types that tend to die out quickly. Certainly in nature many incipient species evolve without expanding their populations or surviving for millions of years. The differential success of small isolates can represent selection at the level of the population.

THE CORRELATION BETWEEN RATES OF SPECIATION AND RATES OF EXTINCTION

Many taxa, when they originate, expand their numbers of species rapidly. An adaptive radiation of this type may result from the origin of the taxon by way of an evolutionary innovation which allows it to diversify without severe impedence from preexisting taxa. Adaptive radiations also frequently follow major extinctions, which create empty ecospace, or the origin of a new habitat that is ripe for colonization.

Unbridled adaptive radiation follows an approximately exponential course. Certainly, in time, radiation will be damped by such factors as the onset of competition between species and the appearance of predatory groups that are specially adapted to attack species formed by the adaptive radiation. For the early stages of adaptive radiation, however, we can employ the following basic equation that describes exponential (or geometric) growth:

$$N = N_o e^{Rt}$$

where N = the final number of species, N_o = the original number of species (one, for monophyly), e = the base of natural (Naperian) logarithms, t = the time over which radiation has taken place (in millions of years), and R = the rate of adaptive radiation (fractional increase per million years). The best place to apply this equation in order to estimate rate of adaptive radiation is the Neogene, where we can employ the modern world as an end point (where N is the number of living species within a radiating taxon). This allows us to obtain a very good estimate of the number of species, whereas for ancient times, because of the incompleteness of the fossil record, we can make only crude estimates. If we know the number of living species within a taxon that today is in the early stages of adaptive radiation, and we know when the radiation began in geologic time so that we have a good estimate of it, then we can calculate R, which is the net rate of radiation. This can be viewed as a fractional, or percentage, increase per unit time. This net increase is not, however, rate of speciation. Rather it represents the difference between rate of speciation and rate of extinction. We can calculate rate of extinction as the reciprocal mean species duration within a taxon. When we make this estimate and add it to R, we obtain an estimate of the rate of speciation in the adaptive radiation we are evaluating.

It is a remarkable fact that when we estimate rates of speciation and rates of extinction during adaptive radiation for a variety of taxa, we find that the two rates are correlated (Stanley 1979) (Fig. 2). In other words, taxa that enjoy high rates of speciation also suffer high rates of extinction. Examples of such taxa are mammals, ammonoids, and trilobites. On the other hand, groups like

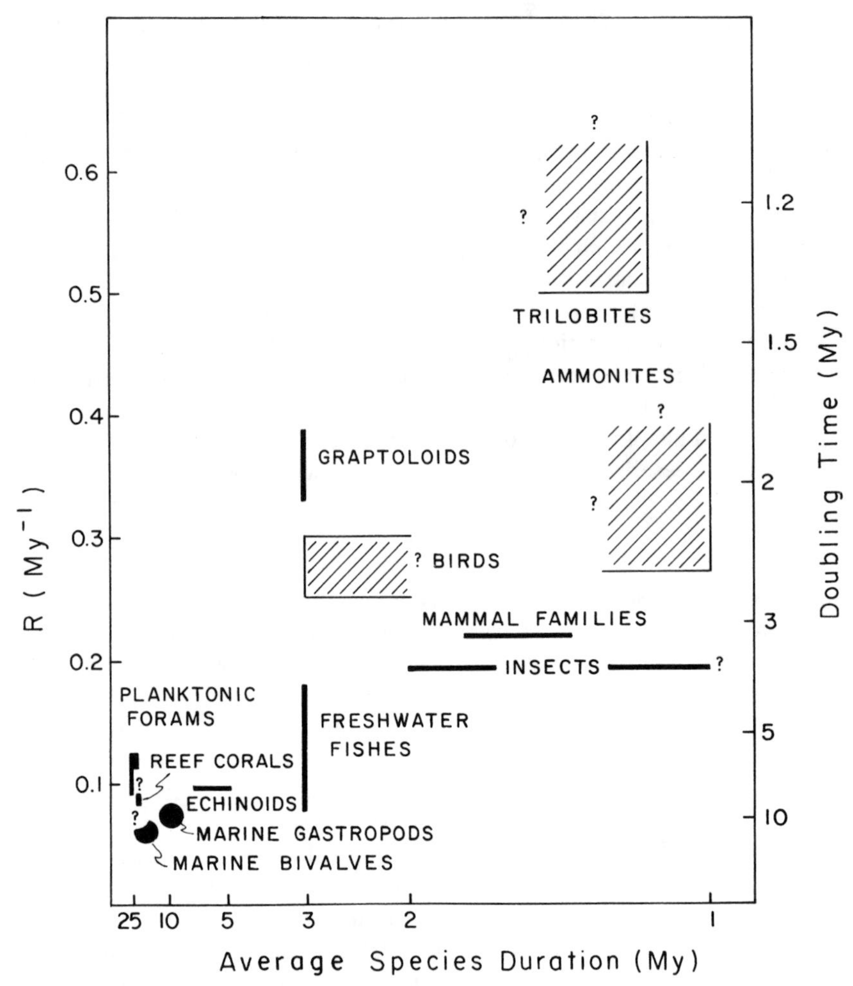

Fig. 2. Plot showing how, for a variety of animal taxa, rate of speciation in adaptive radiation is correlated with rate of extinction. Mean rate of extinction (E) is the inverse of mean species duration. Average species duration is plotted on an inverse scale here, as a surrogate for E, because it is not a precise mean and because pseudoextinction (extinction by gradual transformation) has not been factored out; thus, the position of a group along the horizontal axis is only a rough estimate of mean rate of extinction. R is the measured rate of exponential increase in the number of species for a sizeable taxon early in adaptive between S and E). Even though numbers on the graph are not precise, there are such great differences between taxa with high rates and taxa with low rates tha the trend of the plotted data must be generally accurate. Small pulses of radiation (ones yielding only a few species) can occur within various taxa at higher values of R than are shown here; the rates plotted here represent radiations that have yielded dozens of species (from Stanley 1979).

foraminifera and marine bivalve mollusks have experienced much lower rates of speciation and extinction.

The correlation between rates of speciation and rates of extinction in adaptive radiation requires an explanation. More generally, it is easy to see that taxa cannot have high rates of extinction and low rates of speciation or they would suffer rapid decline to extinction. Not so readily explained is the fact that we find very few taxa in nature that are typified by very high rates of speciation in adaptive radiation and very low rates of extinction. The absence of such taxa is puzzling because species selection, by definition, favors species characterized by high rates of speciation and low rates of extinction. Thus, species selection must always tend to produce higher taxa that are characterized by such rates. We must suspect that the reason that taxa of this type are rare is that rates of speciation and rates of extinction are governed by the same factors—and that these factors influence both kinds of rates in the same way. In other words, if a particular biological trait within a higher taxon tends to promote speciation, it also fortuitously increases rate of extinction.

One variable that may contribute to the general correlation between rates of speciation and rates of extinction is dispersal ability. It appears, for example, that weak dispersal in marine gastropods that lack planktonic larvae results in narrow geographic ranges for individual species, which elevates probability of extinction. This trait, however, also increases the probability that small populations will become isolated and evolve into new species. On the other hand, species having planktonic larvae, which can spread over broad areas, tend to have widespread populations that make these species resistant to extinction; these species also have a reduced probability of producing isolated populations that may evolve into new species. This kind of pattern has been documented for members of the gastropod family Volutidae (Hansen 1980).

Still another variable that seems to operate in the same way is level of advanced behavior. Complex, rigid behavior, while it may represent a useful adaptation in a stable environment, makes a species vulnerable to extinction when the environment changes. Thus, complex, specialized behavior can be expected to be associated with a high rate of extinction. At the same time, this kind of behavior tends to promote speciation because a minor change in complex behavior can easily lead to reproductive isolation and speciation. The generally high rates of speciation and extinction that mammals have experienced can perhaps be attributed in part to the complex behavior of this advanced group of animals.

A third variable that may contribute to the correlation between rates of adaptive radiation and rates of extinction is niche breadth. Narrowly adapted species certainly may be expected to be especially vulnerable to environmental fluctuations and, thus, to be characterized by high average rates of extinction. At the same time, these kinds of species should experience greater opportunities for speciation than broadly adapted species because they are not monopolizing as much ecospace (Slobodkin and Sanders 1969, Jackson 1974) or because they are more frequently subject to directional selection pressure (Vrba 1984).

Species characterized by small unstable populations also appear to experience high rates of both speciation and extinction. Such species are obviously more vulnerable to extinction than are species with large stable populations, but at the same time their patchy distribution makes them susceptible to speciation by the spatial isolation of small populations (Stanley 1986, 1988). In fact, this kind of pattern has been observed for both bivalve mollusks and planktonic foraminifera. Among the bivalves, burrowing marine species that lack siphons have experienced much higher rates of both extinction and speciation during the past few million years than have species which possess siphons. The non-siphonate groups, being slower and shallower burrowers, are more susceptible to predators, and their populations are smaller and less stable. Of the planktonic foraminifera that have inhabited the world's oceans during the past 20 million years, one of the two large groups, the globorotaliids, has experienced higher rates of both speciation and extinction than the other large group, the globigerinids (Stanley et al. 1988). The globigerinid species tend to live in the photic zone, where food is relatively abundant and where many of their representatives employ symbiotic algae to supplement their nutrition. Thus, the globigerinid species tend to have relatively large, stable populations. In contrast, species of the globorotaliid group tend to live deeper in the water column, where food is scarcer and symbiotic algae cannot be employed. Their populations tend to be smaller and less stable; globorotaliids flourish best in local areas where upwelling brings cool water into photic zone. The small size and patchiness of globorotaliid populations appears to endow them with both high rates of extinction and high rates of speciation.

In summary, it appears that the general correlation between rates of speciation in adaptive radiation and rates of extinction exists because the biological traits that tend to promote speciation also happen to make species more vulnerable to extinction. This correlation exists not only in the animal world (Stanley 1979) but also in the plant world, where certain groups of plants in the course of geologic time have experienced much higher rates of evolutionary turnover than others (Niklas et al. 1983).

REFERENCES

Cheetham AH (1986): Tempo of evolution in a Neogene bryozoan: Rates of morphologic change within and across species boundaries. Paleobiol 12:190.

Cronin TM (1987): Evolution, biogeography, and systematics of *Puriana:* Evolution and speciation in Ostracoda, III. Paleont Soc Mem 21.

Eldredge N (1971): The allopatric model and phylogeny in Paleozoic invertebrates. Evol 25:156.

Eldredge N, Gould S J (1972): Punctuated equilibria: An alternative to phyletic gradualism. In Schopf TJM (ed): "Models in Paleobiology." San Francisco: Freeman, Cooper, p 82.

Felsenstein J (1974): The evolutionary advantage of recombination. Genetics 78:737.
Hansen TA (1980): Influence of larval dispersal and geographic distribution on species longevity in neogastropods. Paleobiol 6:193.
Jackson JBC (1974): Biogeographic consequences of eurytopy and stenotopy among marine bivalves and their evolutionary significance. Amer Nat 108:541.
Malmgren BA, Berggren WA, Lohman GP (1983): Evidence for punctuated gradualism in the late Neogene *Globorotalia tumida* lineage of planktonic foraminifera. Paleobiol 10: 377.
Maynard-Smith J (1978): "The Evolution of Sex." Cambridge: Cambridge Univ Press.
Mayr E (1954): Change of genetic environment and evolution. In Huxley J, Hardy AC, Ford EB (eds): "Evolution as a Process." London: Allen & Unwin, p 157.
Mayr E (1971): "Populations, Species, and Evolution." Cambridge (MA): Harvard Univ Press, p 373–374.
Muller HJ (1964): The relation of recombination to mutational advance. Mutat Res 1:2.
Niklas KJ, Tiffney BH, Knoll AH (1983): Patterns in vascular land plant diversification. Nature 303:614.
Nunney L (1989): The maintenance of sex by group selection. Evol 43:245.
Sheldon PR (1987): Parallel gradualistic evolution of Ordovician trilobites. Nature 330:561.
Simpson GG (1951): "Principles of Animal Taxonomy." New York: Columbia Univ Press, p 153.
Simpson GG (1953): "The Major Features of Evolution." New York: Columbia Univ Press, p 31.
Slobodkin LB, Sanders H L (1969): On the contribution of environmental predictability to species diversity. Brookhaven Symp Biol 22:82
Stanley SM (1975a): A theory of evolution above the species level. Proc Nat Acad Sci (USA) 72:646.
Stanley SM (1975b): Clades versus clones in evolution: Why we have sex. Science 190:382.
Stanley, SM (1979): "Macroevolution: Pattern and Process". San Francisco: WH Freeman & Co.
Stanley SM (1981): Species selection involving alternative character states: An approach to macroevolutionary analyses. Third N Amer Paleont Convent Proc 2:505.
Stanley SM (1986): Population size, extinction, and speciation: The fission effect in Neogene *Bivalvia*. Paleobiol 12:89.
Stanley SM, Wetmore KL, Kennett JP (1988): Macroevolutionary differences between the two major clades of Neogene planktonic foraminifera. Paleobiol 14:235.
Stanley SM, Yang X (1987): Approximate evolutionary stasis for bivalve morphology over millions of years: A multivariate, multilineage study. Paleobiol 13:113.
Vrba, ES (1984): Evolutionary patterns and process in the sister-group Alcelaphini-Aepycerotini (Mammalia: Bovidae). In Eldredge N, Stanley SM (eds): "Living Fossils." New York: Springler-Verlag, pp 62–79.
Wright S (1956): Modes of Selection. Amer Nat 90:5.

New Perspectives on Evolution, pages 101-122

Parallelism and Convergence in the Horse Limb: The Internal-External Dichotomy

KEITH STEWART THOMSON
Academy of Natural Sciences, Philadelphia, Pennsylvania 19103

INTRODUCTION

Science is a process of organizing knowledge according to causes. Because biology is quintessentially a comparative science, organizing the fundamental data of biology—our knowledge of the organisms themselves—according to cause is essential to biology. That is to say, evolution and systematics are the foundations of biology. Although molecular biology provides (or will eventually provide) a universal language for the discussion of mechanistic causes, not all causal factors can be reduced solely to processes acting at the molecular level.

"The whole guts of evolution—which is, how do you come to have horses and tigers. . . ." (Waddington, 1967a) is a question that often gets lost in the detailed exposition of a particular theory or the collection of data around a particular experimental approach. What follows is an essay on the role of development in the evolution of morphology and the longstanding questions concerning how complex adaptations arise not only once, which is unlikely enough, but more than once either in parallel or convergently.

We must begin by stating the obvious: The term *evolution* means different things to different people. The fundamentals of the general concept of organic evolution are (1) change over time and (2) relationship of organisms by descent from common ancestry. We owe it to Darwin that the term evolution commonly also means (3) a particular mechanism—natural selection—and the study of evolution has slowly shifted from the documentation of change and of relationship to the study of cause. The search for evidence on the causes of evolution, which means the search for data to support or deny given theories, still has not produced the smoking gun, to use the metaphor beloved of students of political intrigue. Not surprisingly, therefore, different people study evolution in quite different frames of reference. And, equally unsurprisingly, the study of evolutionary mechanisms suffers as much from the vagaries of fashion and the domination of personalities as any other wholly human endeavour.

We can mark the course of the study of evolution by milestones represented by the publication of a small number of significant works. In this century there is a particular importance to Fisher's *Genetical Theory of Natural Selection* (1930), Dobzhansky's *Genetics and the Origin of Species* (1937), Huxley's (1942) *Evolution: The Modern Synthesis;* Simpson's *Tempo and Mode in Evolution* (1942), Jepsen, Mayr and Simpson's *Genetics, Paleontology and Evolution* (1949), and Mayr's *Animal Species and Evolution* (1963). These works shaped and their dates bracket a phase in the study of evolutionary mechanisms and theory that can be called "externalist." The point-counterpoint of Fisher and Wright marked a reaction to an older, distinctly internalist, approach of the sort that no-one follows now: Relics of the old ontogeny/phylogeny story and in particular the work of Bateson and others on directed mutation. Along the way the New Synthetic Theory of Evolution swept aside the work of people like Goldschmidt (the good with the bad) and relegated to the status of interesting curiosity studies like that of de Beer (1958) on heterochrony or Waddington's developmental genetics (e.g., 1967b). In the wake of this, paleontology sought to find intellectual respectability in microevolutionary terms. Developmental approaches to evolution dwindled to nothing (and indeed, at around this time, lacking an effective set of new technologies, developmental biology as a whole declined).

In this paper I will try to show that the wholly externalist approach to evolutionary mechanisms of the New Synthesis, while valid and important within its own frame of reference, is not the whole answer to the problem, and that an internalist approach—one far different from antique directed mutation or hopeful monster theories—must be added to the hierarchy of causes of evolution. I will do so by examining the age-old problem of parallelism and convergence in evolution, and the related phenomenon of trends. These are subjects that a purely externalist approach has so far failed to solve and indeed has in recent years significantly avoided addressing. Even a cursory examination of parallelism, convergence and trends, however, forces one to see the need for an internalist approach.

EXTERNAL, INTERNAL, AND HIERARCHICAL

An externalist approach is one that seeks to explain the rate and direction (the mode) of evolutionary mechanisms solely by the operation of a gamut of selection mechanisms. The source of the variation upon which these mechanisms operate is then essentially irrelevant because variation is *only* [sic] randomly generated with respect to the intensity and direction of selection. While no modern biologist believes that variation expressed at the level of the individual organism (or the deme for that matter) is a direct representation of variation occurring at the genome level, nonetheless purely externalist approaches do not incorporate (or indeed require) any reference to the black box that links genetic to phenotypic variation. Most people are quite happy with a vague reductionist approach that believes that the study of molecular genetics will one day turn out to be extremely important to the study of selection, but as no-one can see just how the link would be made, the internal causes of variation can be avoided, or sublimated by studying genetic information as just another sort of character.

A modern internalist approach does not deny the importance of selection, including strong directed selection, in evolutionary mechanisms, but it places at least as much emphasis on the processes that create the variation in the first place. For obviously one cannot select what has not already been created. Indeed, somewhat paradoxically, while it is a sacred given that selection cannot change the direction of variation (although there are some interesting chinks in this premise), the reverse is not often given enough weight: the nature of variation must in part direct the course of selection. This means that any new variant changes the rules by creating its own subtly different environment in which to be tested, rather than being tested solely in terms of the parameters in which the original was tested (Thomson 1988a).

An extreme externalist view would be that, given enough time, strong directed selection could produce almost any morphology in any lineage. To be sure, the course of evolution has produced some superbly unlikely morphologies, but no-one seriously believes the above statement to be true. We implicitly accept that certain lineages have properties that make certain evolutionary eventualities more likely than others. Even if we only accept the simplest notions of phyletic constraints or developmental constraints (see Smith et al. 1985; Thomson 1986, 1988b), we are already part way to accepting the validity of internalist approaches to evolutionary theory.

An internalist approach says that the properties of the system that generates the variation are at least as interesting and potentially as important in shaping the course of evolution as the properties of the selection mechanism working on that variation. The particular nature of the developing pathways and the ways in which they respond to perturbation may be significant contributing factors in evolution. In fact, for a long time people have been aware of this at the population level. In population genetics, breeding systems, population size, and other rules-of-assembly factors, internal to the deme or population in question, are as important as the selective regime. Indeed, in special cases like the founder effect or genetic drift, populations may produce change essentially independent of selection. If this is true for demes or populations, and if it might also be true for species (see the whole debate over species selection), then is it true for other mechanistic systems as well? This brings us to hierarchy theory.

The fundamentals of any evolutionary mechanism are (1) the causation and introduction of variation and (2) the sorting (including selection) of that variation. The basis of using hierarchy theory as an approach to understanding evolutionary mechanisms (Vrba and Eldredge 1984) is the premise that the dual processes of introduction and sorting of variation occur at multiple hierarchical levels. In mechanisms of change by descent in lineages of organisms, the causes of evolutionary change must exist, and therefore can be studied, at least at the following levels: the genome, the individual organism, the deme, the species and perhaps beyond (Fig. 1). (Of course, other possible hierarchies exist. If we were to create an ecological hierarchy we could look at evolution at an angle roughly rectilinear to this one.)

Just as there is a complex set of causal factors that translate the information contained in a large number of individual organisms into the properties of a deme or a population, so there is a vast set of causal mechanisms staged between the organization and function of the genome and the information contained in individual organisms. That black box is easily labeled—development. While the externalist approaches of the New Synthesis concentrate on mechanisms of sorting at the individual organism and population levels, an internalist approach to evolutionary mechanisms is one that pays strong attention to the operation of the black box (really a series of black boxes) that gives us the phenotypic characters upon which natural selection can operate at the level of the individual organism and above. A complete theory of evolution must account for introduction and sorting mechanisms at all hierarchical levels. In one simple step, hierarchical analysis shows the importance of internalist approaches and their relationship to externalist ones.

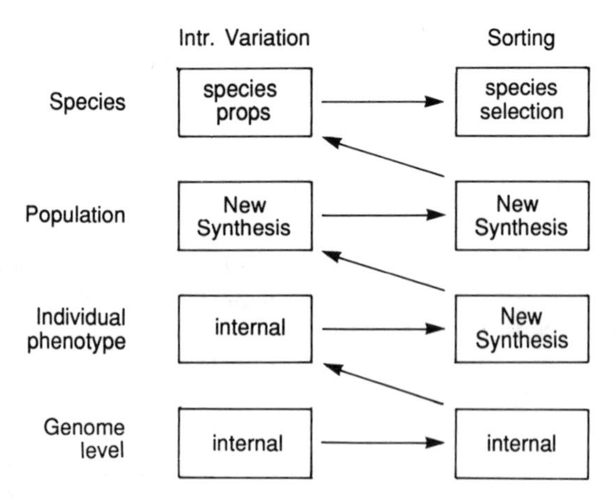

Fig. 1. Basic hierarchical analysis of evolutionary mechanisms (after Vrba and Eldredge). At each of the four focal levels, processes of introduction of variation and sorting of variation operate. The realms conventionally ascribed to the (externalist) New Synthetic theory, species selection, and internalist mechanisms are indicated.

PARALLELISM AND CONVERGENCE

In any elementary textbook of evolution, a significant amount of space is given to examples of parallelism and convergence. Ever since Huxley and Marsh, one of the great examples of evolution in action has been the evolution of the horse limb with its dreaded overtones of direction, trends, and Osbornian orthogenesis. Curiously, however, in the same textbooks, and in all research in evolution, one finds a curious omission. No mechanistic explanation is ever offered for the cause of these phenomena that are among the most important prima facie evidence that evolution has been going on.

This reticence of evolutionists to explain the most powerful examples of evolution is an example of what I have termed (Thomson 1988a) "the gradualists' dilemma"—how to explain the magnificently improbable from the ineffably trivial. Yet these are the sorts of questions that must be faced, because a weakness like this is readily exploitable by those opposed to theories of evolution.

The case that I wish to consider here is a classic case of parallelism and convergence (for definitions, see Fig. 2) in the evolution of the limbs of artiodactyls and perissodactyls, whose limbs are so wonderfully adapted for fast running. Parallelism occurs in the evolution of an unguligrade limb with elongated and reduced manus and pes in both artiodactyls and perissodactyls. There is parallelism in the multiple origins of both three-toed horses in the Miocene and one-toed horses in the Pliocene. Convergence occurs in the separate evolution of a one-toed foot in both the horses and the South American notungulate order Litopterna (Fig. 3). Apparent trends occur in all three sets of radiations.

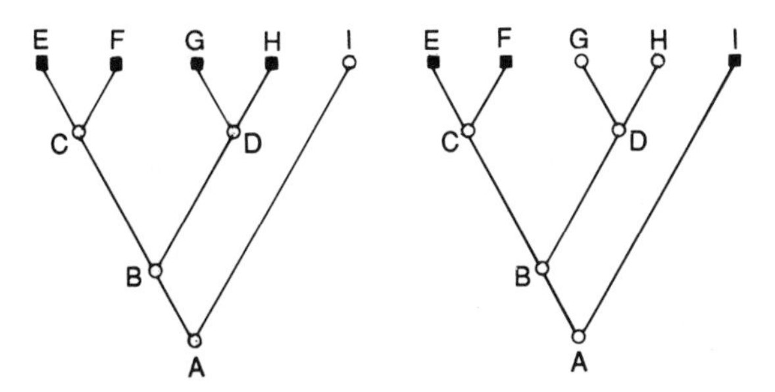

Fig. 2. Cladograms showing the patterns of origin of characters that can be defined as parallelism **(left)** and convergence **(right).** The black squares indicate the newly arisen character state. In the left hand figure, parallelism occurs in the origin of the derived character state in **E, F, G,** and **H** from a common ancestor **B.** The fact that the derived character is not present in **C** or **D** shows that the character has indeed arisen more than once in parallel. In the right hand figure the derived character states in **E, F,** and **I** do not share a recent common ancestor and so the pattern is called convergence.

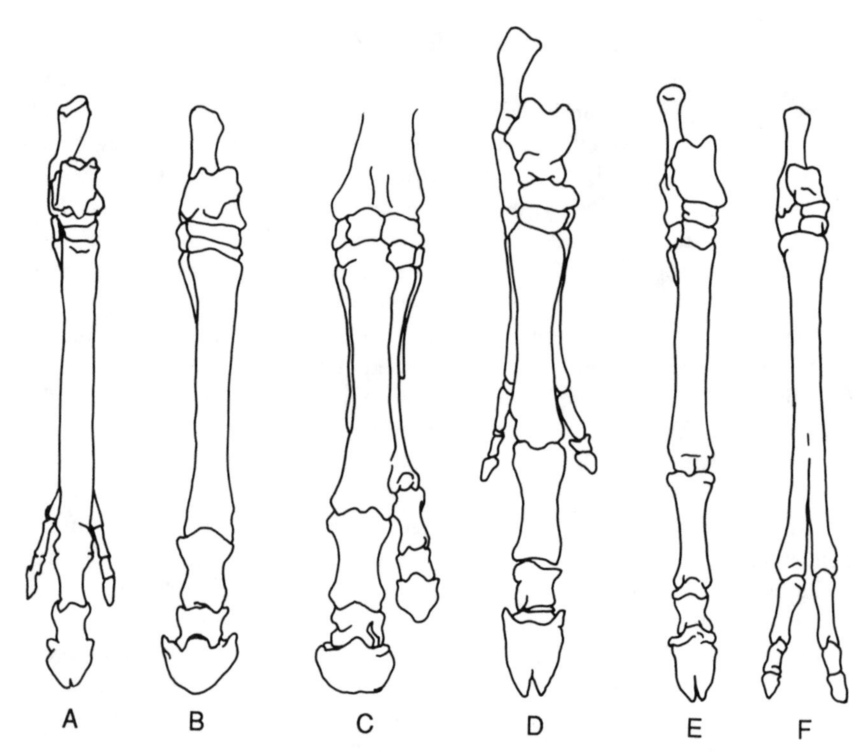

Fig. 3. Structure of the foot in some ungulates. **A:** the Miocene horse *Merychippus,* **B:** modern *Equus,* **C:** a polydactyl *Equus* with an atavistic second digit, **D:** the Miocene litoptern *Diadiaphorus,* **E:** the Miocene litoptern *Thoatherium,* **F:** the Pliocene camel *Procamelus* (after Romer and Marsh).

A standard explanation of these adaptations is that they represent the functionally determined response to a common set of environmental constraints—in this case, the need for fast running over hard grasslands. They have occurred through the slow but steady directional selection of randomly occurring variants of the same general sort as occur in any mammalian population. There is no inherent tendency of ungulates towards lateral digit reduction. As Smith puts it: "The main trends in horse evolution can reasonably be explained as adaptations to a herbivorous life, produced by natural selection. When a change of habits occurred among grazing horses, this was accompanied by rapid changes in the structure of teeth and feet. In so far as a number of related lineages tend to undergo similar evolutionary changes, this is because groups of animals which had adopted similar ways of life are subjected to similar selection pressures" (1966:266). This is an externalist explanation: The key words here, however, are *change of habits* (that is, which is the chicken and which is the egg?), *rapid, similar,* and *related.*

Parallelism and convergence then had to be explained in the same externalist terms, and particularly ones that do not require the origin of identical genetic variations in different lineages. Thus, "if there is only one efficient solution for a certain functional demand, very different gene complexes will come up with the same solution" (Mayr 1963). This expresses the common view of the extreme opportunism of the genome and plasticity of phenotype. The term to worry about here is *one efficient solution.* It is worth noting that mammals run fast in lots of ways. For example, none of the predators from which horses presumably are adapted to escape with their wonderful running (they do not eat on the run, after all) have unguligrade locomotion. Neither do kangaroos, although they do have spring loading of the foot. This serves to remind us not to fall into the trap of trying to explain horse evolution via selection *for* unguligrady but rather by selection *of* unguligrady. And that distinction forces us to concentrate on where the unguligrady came from—i.e., the causes of the variation concerned.

LIMB MODIFICATION IN UNGULATES

As is well-known, the modern horse genus *Equus* and its close Pliocene and Pleistocene relatives are one-toed horses with an extreme unguligrade locomotion (Fig. 3). The fore and hind limbs are elongated and the manus and pes are especially important elements of the limb lever system. The humerus and femur are relatively short. The ulna and fibula are strongly reduced and lack separate distal articular epiphyses. In the feet there is a single strong cannon bone, which is the middle or third metapodial, plus a splint bone which is a reduced fourth metapodial. An important element of the running mechanism is the spring ligament system (Camp and Smith 1942).

Less well-known is the fact that an extremely similar monopodial limb also occurs in the distantly related South American ungulates of the Order Litopterna (Figure 3E), although here the adaptation of the carpus and tarsus is not as advanced as in horses. The closest common ancestor shared by horses and litopterns seems to be the Paleocene Condylartha, which were five-toed animals. Whether this is parallelism or convergence depends entirely upon the level of taxonomic comparison.

Within the horses, however, there was by no means a simple unidirectional trend towards the development of the monodactyl limb. In the transition from the Eocene *Hyracotherium,* with four toes in the manus and three in the pes, to the three-toed Miocene horses like *Miohippus* and *Merychippus,* reduction of digit four in the manus probably occurred several times independently. Similarly, the evolution of the single-toed condition seems to have occurred in parallel more than once among the Miocene/Pliocene descendants of the three-toed forms (review in MacFadden 1988). The fossil record shows that the characteristic spring-ligament mechanism of the modern horse foot arose from within the late Miocene radiation of three-toed horses (Simpson 1961).

The general functional sister group of the horses is the Artiodactyla (also condylarth descendants) and here we find an interesting case of parallelism in that once again the limbs are strongly modified for unguligrade locomotion. But instead of a mesaxonic monopodial foot evolving, we have the paraxonic cloven two- or four-digit foot. Camels afford an example of extreme limb reduction and are functionally analogous to the highly cursorial horses, including an analogous spring loading of the foot. The ulna and fibula are very strongly reduced and metapodials three and four are fused to form a cannon bone superficially quite similar to that of horses except that it supports two toes (Figure 3F). In camels no other metapodials exist, even as splints. In other less highly modified ungulates such as pigs, toes two and five are functional, although smaller than three and four; the metapodials are not completely fused to form a cannon bone.

What can paleontology tell us about the historical course of limb changes in ungulate evolution? First, of course, changes in the limbs do not occur in isolation. In both Perissodactyla and Artiodactyla, the whole morphotype changes. Particularly, there is a comparable set of changes in the head and dentition that has its own parallelism and convergence—in concert with the limb changes. Interestingly, modification of the head and dentition in the Litopterna is much less pronounced.

A lot of attention has been given to the role of size changes in all of this, particularly with a view to explaining it all in terms of allometry and heterochrony. At least as far as horse evolution is concerned, however, Radinsky (1984) and MacFadden (1986) have shown that there are many significant effects that are not due to allometry. In fact, one can roughly divide horse evolution into two phases. The first phase, from *Hyracotherium* to the Miocene three-toed horses is largely passed through without major size increase. This is a phase of change of proportions of the limbs. The second phase, the Miocene three-toed horses to the Pliocene and Modern single-toed horses, is one of dramatic size increase (and also, at least one lineage, dwarfism). The fossil record shows considerable variability in the number and relative size of lateral digits among the Miocene three-toed horses (Prothero and Shubin 1986).

DEVELOPMENTAL PATHWAYS AND THE CONTROL OF PHENOTYPIC CHARACTERS

In general it is useful to consider the course of development of an organism (or the developmental pathway for any single feature) as a series of cascades of decision points (discussion in Thomson 1988b). At each decision point gene expression is controlled. Every new element of gene expression then goes to build the environment that will control the next stage of gene expression.

Almost nothing that happens in these cascades of gene expression occurs independently of what is going on in other pathways, especially early in development. A simple model of pattern control in a developmental pathway can be built up according to what can be called the "developmental rules" of particular mechanisms—for example, folding of an epithelium, reaction of a group of cells to an ECM, growth of a blastema, the rate of a reaction. For each rule there are boundary and initial conditions. The result of gene expression at level A in the cascade produces new gene expression B, which then helps define the set of initial conditions for a second function B with its own rules, boundary and initial conditions, and so on (Fig. 4). At each stage the initial conditions are also dictated by all the functioning of *adjoining* pathways (geographical terms like adjoining are not quite correct here but the metaphor is understandable).

The control of morphogenesis or any other element of development is therefore under multiple control from the specifics of gene expression and the generalities of the epigenetic environment in which any aspect of development and the development of the embryo as a whole, are occurring. Indeed, many of the crucial features of developmental pathways are epigenetic rather than genetic. The result (as Waddington so eloquently argued, especially with his canalization metaphor, 1967b) is that development as a whole and many sub-components of development, especially early in morphogenesis, are buffered and self-regulating in the face of perturbation (Rachootin and Thomson 1981; Thomson 1986, 1988b). The result also is that changes to developmental systems are likely to be threshold effects.

Within these cascades lie the control of phenotypic characters and phenotypic variation (pattern control). There is a contribution to the cause of phenotypic variation at each and every stage of development from fertilization of the egg to final cytodifferentiation. But a crucial question is, while minor variations can be caused at any level of development, where are large-scale phenotypic

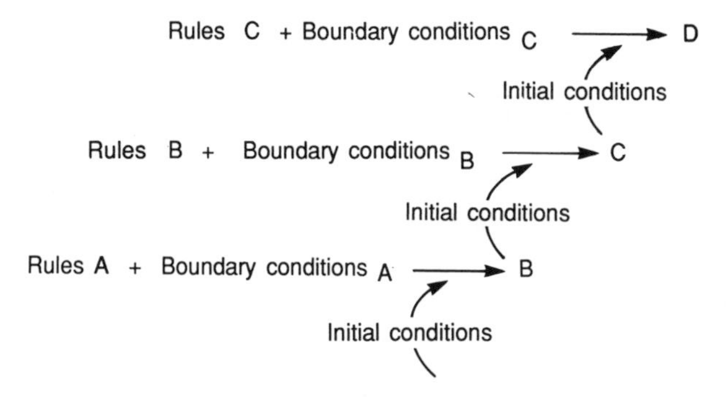

Fig. 4. Hierarchical nature of the pattern control process in morphogenesis, showing three hypothetical processes, their rules, boundary conditions, initial conditions, and interactions.

differences among lineages caused? Specifically, what is the site of the changes in developmental pathways that account for major morphological changes such as loss of lateral digits?

Because of the essentially holistic nature of development and the multiplicative properties of developmental cascades, logically it should be the case that a major phenotypic difference between two lineages should be caused by extremely subtle changes first expressed relatively early in development. A subtle change would cause subtle effects in intersecting patterns with the result that the whole would be accommodated, producing a viable embryo. Equally, at a range of later stages in development, when morphogenesis is further advanced and the capacity for buffering or regulation proportionately reduced, such a shift would cause major disruption of pathways. Then, towards the end of development, each pathway is much more independent and variation is tolerable again, but the opportunity for causing major change is commensurately reduced.

This matter is important because one of the crucial arguments against any internalist developmental explanation of causality in evolution has always been the premise that any change to early development would be so disruptive to so many systems that it would inevitably be lethal. The first thing to say about this premise is that obviously early developmental stages can be and have been modified because, if one compares organisms of different lineages, they differ at early as well as late developmental levels. The idea of Muller that the earlier you trace into development the more embryos are the same is obviously wrong. One needs to consider the matter of scale. As you go deeper you have to look at a finer grain—at molecular rather than gross tissue and cell levels. This produces a problem that Haeckel tried to address. If early development does not change, then evolution must proceed by accumulation of changes late in development—so-called terminal additions. But then, at some point or other, developmental patterns had to have been secondarily reorganized to produce the gradually and subtly divergent patterns that we actually observe. Muller by contrast seems to have recognized what is now obvious, that it is simpler to assume that evolution proceeded by changes at all levels of development including early stages (Lovtrup 1987).

Whether we are discussing morphogenesis or the development and evolution of metabolic pathways, in the ontogeny of a single organism, minor variations introduced early in development might have one of several results, both over ontogenetic or evolutionary time scales. (1) They could be so buffered as to fail to modify the phenotype. (2) They might produce a subtle change in phenotype. (3) Their effects might become multiplied within the cascades of development so as to produce a more significant change in phenotype. It is always argued, of course, that in the last case, chances for deleterious or lethal change are increased. Even if this is true, the effect will still be less if the effect arises early enough, because of the integrative nature of the developmental processes.

Then, as lineages diverge, if the cause of divergence is a change that is set in place early in development, the greater the chance will be for coherent change to accumulate. Conversely, while it is usually argued that late developmental changes will be less disruptive to the phenotype (and this may well be true for single minor changes), such changes cannot accumulate to major changes without disruption of the phenotype because they are put into place too late for the pattern control mechanism to respond by producing a fully integrated phenotype (Thomson 1988b). The reason that this has not been fully appreciated before is possibly that one tends to fall into the mistake of thinking of gene expression producing things instead of controlling processes (see below).

In short, minor phenotypic variation and minor deviations in morphogenetic pathways are probably caused at a whole variety of development levels from early to late. Major deviations among lineages, however, are most likely only set in place early. They probably start out, both in evolutionary and phenotypic terms, as minor variations (there is no need to invoke macromutations) but they have a different evolutionary potential from a minor variant set in place late.

DEVELOPMENT OF THE LIMB IN GENERAL

The general pattern of limb skeleton development in vertebrates is well known, although there are very many details that await elucidation (reviews in Hinchliffe and Johnson 1980; Shubin and Alberch 1986). We can crudely divide the course of development into a series of stages. They are

1. Initial limb bud stage: formation of AER, ZPA, first mesenchyme; axis determination and polarity; gradient and other effects.
2. Early mesenchymal stage: pattern formation continues; formation of first suggestion of discrete blastemata.
3. Blastema stage: growth and branching of blastemata.
4. Chondrification stage.
5. Ossification stage.

Throughout this cascade of events, the phenotype is created from the operation of a finite number of general processes, each with their own rules, boundary conditions and initial conditions. These processes include particularly the growth of mesenchymal rudiments through cell division and recruitment, formation of ECMs and cell response to ECMs, the growth and patterning of muscles and blood vessels, the immigration of nerves. For example, the rate and extent of hyaluronate removal by hyaluronidase from the initial mesenchyme cell masses affects the cell number and patterning of the blastemata (Toole et al. 1972). Cell number and position in turn affect cell shape in the blastemata, and these in turn control both the pattern of further recruitment to the blastema and the pattern of secretion of glycosaminoglycans in the start of cartilage formation (Fig. 5).

It is in the control of these processes that changes in the pattern formation mechanisms of the limb can be effected. There are no discrete and unique codings for the shape of the second phalanx of digit three or the length of the ulna; instead we have processes that affect general quantitative properties of the developing limb, whose control is essentially epigenetic rather than genetic. Thus selection does not, at any stage from the genomic to the phenotypic, act on single character states.

For example, whether a rudiment ossifies or not may depend heavily upon initial size and cell number, and thus depend on the growth rate of the previous chondrification stage, which was affected by its initial cell number as well as the conditions for its growth; these in turn were a result of recruitment and growth phenomena at the mesenchymal stage, which were set in place by basic patterning processes acting at the preblastema stage (Fig. 6).

The cascades of decisions in developmental pathways form a developmental hierarchy (Thomson 1988b). Any change effected at early stages will poten-

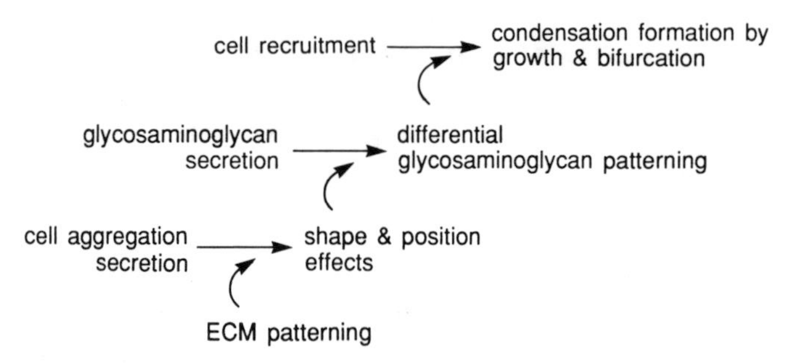

Fig. 5. Three stages in the hierarchy of skeletal morphogenesis and their possible interactions (cf. Fig. 4).

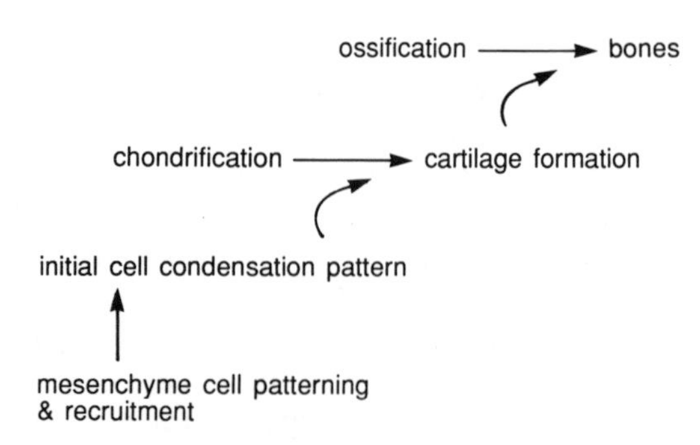

Fig. 6. Hierarchical arrangement of four crucial stages in skeletal morphogenesis at which the developmental cascade may be interrupted.

tially be felt right along the course of development. Because this is a cascade of processes affecting a single whole unit—the limb—rather than an assembly line of successive control over discrete structural elements, subtle quantitative changes effected early in development will have the best chance of producing coordinated non-deleterious variation in the whole limb phenotype. After a certain point in limb development, changes in the system will only produce effects in different discrete structures and then the result has less chance of producing a functionally coordinated whole (Fig. 7). A classic example of the functionally integrated remodelling of the phenotype is Muller's work (1986) with experimentally produced changes in growth rates of skeletal blastemata in the chick wing.

UNGULATE LIMB DEVELOPMENT

"Notwithstanding the fact that, alike from a scientific and a practical standpoint, the limbs of the Horse have long demanded our careful consideration, there does not yet exist anything approaching a complete or accurate account of their development. (Ewart 1894a). Ewart's observation is still true, nearly 100 years later.

We do know that a number of elements that are not represented in the adult phenotype are present at varying earlier developmental stages. Specifically, in the horse, metapodials two and four are usually present early in development (Fig. 8) but metapodial two is usually lost before the ossification stage. At early stages of development (late blastema/early chondrification) the ulna and fibula are relatively well developed and contact the carpus or tarsus, respectively (Ewart 1894a,b).

While we have very little direct information concerning the all-important early blastema and preblastema stages of the early limb bud in the horse, there

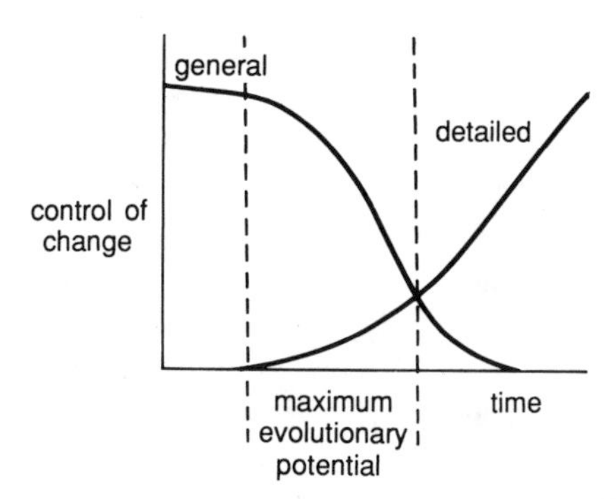

Fig. 7. Idealized model of the time course of relative importance in ontogeny of factors affecting the whole organism *(general)* and any particular structure *(detailed)*. The phase of maximum evolutionary potential is that in which small changes to the developmental program have the best chance of contributing to divergence of lineages without lethality.

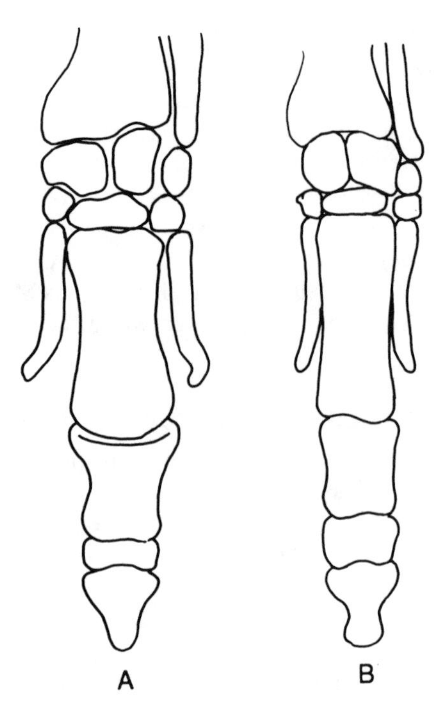

Fig. 8. The arrangement of the foot skeleton of the horse in two very early stages of development. **A:** 20-mm embryo; **B:** 25-mm embryo. Note that both the ulna and radius participate in the "wrist" joint and metapodials two and four are present. (after Ewart; not to scale).

is a little evidence from mutant conditions. Among the mutant conditions described for the horse, the most significant for our purposes are the occurrence of a complete ulna and fibula in the Shetland Pony (Speed 1958; Hermans 1969; Shamis 1985) and the many instances of polydactyly (Fig. 9). Perhaps the first famous polydactyl horse was one bred and ridden by Julius Caesar, as reported by Suetonius. Marsh (1892) and Prentiss (1903) record many other examples, the trait being common according to Marsh in western Mustangs. Both two- and three-toed conditions are reported and Prentiss was at pains to show that many of the cases represented genuine persistence into the adult phenotype of digits two and four, rather than purely secondary division of digit three. Interestingly, the two-toed condition is more common than the three-toed, and the fore feet are more susceptible than the hind. When a second digit is present, however, it is not the fourth (which would seem logical, given the persistence of the fourth metapodial in the wild type condition) but the second, complete with a metapodial that is usually partially fused with the cannon bone (Fig. 3C). Finally, in addition to the occurrence of digits two and four as functional digits (at least to the point of having tiny hooves),

Fig. 9. Three examples of polydactyly in horses. **A:** second digit present in manus and pes; **B:** second digit present in manus only; C: second digit present in manus, second and fourth digits present in pes (after Marsh).

digit five is also frequently present in polydactylous horses in the form of a small metapodial splint with no terminal phalanges. No vestige of digit one seems to be seen.

Given all these atavistic mutant possibilities, it becomes the more important to know more about the normal course of development of the horse limb and attempts are presently under way by this author to remedy this deficiency. Ewart (1894a,b) claimed to have seen, in a 20mm embryo which was in the late blastema/early chondrification stage, not only the blastemata of metapodials two and four, but also nubbins of tissue representing the traces of phalangeal elements of digits two and four. These claims have been disputed by Carlens (1927).

In Artiodactyla, both polydactyly and syndactyly are reasonably common (reviews in Prentiss 1903, Carlens 1927, Johnson et al. 1982). Polydactyly occurs both through genuine atavism and secondary division of digits and primary polydactyly appears to be a multigenic effect. Polydactyly may produce supernumerary digits or merely extra metapodials (two and five). Johnson et al. (1982) theorize that polydactyly is caused by enzyme deficiencies in very early development that disturb the rates of cell division leading to an increased number of mesenchyme cells being formed while the patterning of the blastemata is being controlled.

Gruneberg (1963), Gruneberg and Huston (1965), Leipold (1973) and others have studied syndactyly in cattle extensively. Bovine syndactyly appears to be caused by or triggered by a single autosomal gene *syn.* In normal development, metapodials three and four are separate at the blastema and chondrification stage and therefore fuse rather late in development. By contrast, the syndactylous effect is produced by a fusion of the three and fourth digits at the blastema stage; the process begins distally and proceeds proximally. Gruneberg and Huston (1965) explain the origin of the trait as a threshold phenomena caused by an abnormal proximity of the blastema in the limb bud. This is turn is caused by an abnormally thick and narrow shape of the limb bud. But it is interesting that the mesenchymal blastemata start out separate, which would indicate either that the point of direct expression of the defect occurs after blastema patterning and growth, or else that the limb is able to regulate the effects of the abnormality at the blastema patterning stage but at later stages cannot.

A very important feature of the polydactylous conditions reported in ungulates and other vertebrates is that the arrangement of the soft tissues—nerves, muscles and blood vessels—is often adaptive and functional. This serves to emphasize that we should not concentrate on the skeleton alone but rather must consider the whole limb as a unit. (There is a great tendency to imagine that the soft tissues fall in place around the hard tissues, whereas it could equally be the other way around, and most likely it happens as a whole organ effect.)

While strongly resisting any tendency to analyze the development of the ungulate limb in terms of a recapitulation of evolutionary history, we can look for clues to show us the points in the developmental cascades of limb morphology at which characteristics like lateral digit reduction are controlled and, by extension, must have first been introduced as variants in populations of extinct horses. The mutant evidence shows that the key variants were caused (and control of the modern phenotype now occurs) no later than the blastema stage and probably in the earliest limb bud stages when the shape of the limb bud was fixed and the fundamental pattern control mechanisms first are becoming expressed. The result of these subtle changes in very early limb development is to set in train processes that cause digit development to cease at later stages. Probably crucial to the process were the initial patterning of the mesenchyme through recruitment and growth, initial cell number at each stage (blastema, chondrification, and ossification), relative growth rates of the skeletal and muscle blastemata, the patterning of the vascular system relative to the skeletal blastemata, and the overall shape of the limb bud through ontogeny. All of these are strongly influenced by factors operating at the very earliest stages of limb development. However, it is interesting that the fundamental patterning mechanism of the limb has not been changed from the point in history at which the three-toed foot was fixed. What we see in the one-toed horses and the atavistic mutants is a set of changes in degree of expression of the three-toed system, not a fully fixed one-toed pattern.

Digit reduction is not caused by fixation of alleles that specifically cause, for example, the terminal phalanx of the fourth digit, or the metapodial of the second digit, to abort at, say, the ossification stage and that are expressed at that particular stage in the cascade. Instead what has been fixed is changes in quantitative phenomena set in place by early stage gene expression whose result must be largely controlled epigenetically during the course of further development and can be amplified or overridden by other effects (for example, changing cell division rate and cell number) further down the cascades. All along there are threshold effects, most particularly with respect to critical mass of rudiments at the transition between stages (blastema to chondrification, for example). Hence, many, but not all, phenotypic effects are manifest in a distal-proximal sequence: For example, failure successfully to complete the ossification stage occurs first in the most terminal phalangeal elements, while the metapodials may persist.

A MODEL OF THE ORIGIN OF UNGULIGRADY

Any explanation of the evolution of complex morphological adaptation has to include both selection or other sorting mechanisms and variation. As noted in the Introduction, the relationship between these two is usually thought of as unidirectional: Selection acts on variation. However, quite obviously, neither the quantity of selection pressure nor the direction of selection is truly independent of the nature of the organism. No selection exists without the organism. Selection, both its quantity and its direction, can only be measured after the fact, in terms of organismal responses (individual organism or lineage).

The interplay between selection and variation is at the least an iterative process in which both elements can in principle play an active role. The rate of change is a function of both the rate of variation and the intensity of selection. An internalist approach adds the consideration that the direction of change may also depend in great part upon the nature of the initial adaptation and the direction of variation from that phenotypic condition because they together define the realm in which selection will operate.

1. The paleontological evidence crudely divides horse limb evolution into the following two stages: (A) change of proportions of the limb and loss of digits one and five (five before one; one lost earlier in the pes than the manus); followed by (B) significant increase in size and loss of digits two and four. The first step in the hypothetical model is selection of variants favoring improved fast locomotion. Here there really is universally one efficient solution—namely the change of proportions so that the proximal limb segment is proportionately shorter and the distal segments are longer, especially with elongation of the manus and pes as a third lever element in the limb.
2. The genetic changes that are fixed to produce changes in limb proportion probably were first those that affected relative growth rates. However, the scale and the rate of evolutionary change achieved through changes in growth rate are limited. Further response to strong directional selection involved more fundamental changes affecting, for example, the basic pattern control of the limb, the initial distribution of mesenchyme to the blastemata, competitive recruitment of mesenchyme to the rudiments, as well as the relative growth rates of skeletal (and related) rudiments at all stages of development.
3. Under strong directional selection, the developmental system was not capable of maintaining critical quantitative relationships at crucial stages in all mesenchymal rudiments. The impacted systems probably included the distribution of ECMs during the formation of mesenchymal blastemata, and initial cell size and cell number of mesenchymal rudiments. What the actual mechanistic causes and their controls are is still not understood in any limb system, but the result here was a series of threshold effects as skeletal rudiments failed to complete the developmental cascade. These threshold phenomena affected, for example, the capacity of elements to pass from the blastema to chondrification, or from chondrification to ossification stage. The results were be manifest differentially in the transverse and proximo-distal axis of the limb, so that roughly

speaking the last elements to form in sequence at the early blastema stage were the first elements to be affected (see, for example, Alberch and Gale 1983; Shubin and Alberch 1986). As digit one seems never to be represented in any living or fossil horse, we may guess that it was lost first, followed by number five, and at the same time this was accompanied by strong variability with respect to the size and completeness of digits two and four (in Miocene horses).

4. In addition to this type of internal response of limb development to produce a modified limb phenotype in response to selection for change in limb proportions, the variant phenotypes with small complete second and third digits were also (quite independently) positively selected for, because they also happened to function effectively as part of an efficient design of the foot for fast running. Now probably the lateral digits started to act in part as a check against hyperextension of the foot. At this stage, in Miocene horses, there may well still have been mesenchymal and even chondrified rudiments of digits one and five in the developing foot.
5. In the redesigned three-toed, elongated foot, the check ligaments became especially important and sometime in the Miocene evolved from a passive to a spring-loaded system. Probably this structurally minor but functionally major shift in the properties of the ligament in part allowed the shift of horse lineages to a significantly increased size range. Whatever the functional shift, at this point, variants at the large end of the size spectrum, instead of being selected against (because they were less efficient under the old three-toed functional system) became favorable.
6. All the preceding changes set in place a set of developmental pathways that could not respond to the next phase of selection—for major size increase—without further change. However, the nature and direction of these new changes was already set in place, because the epigenotype (Rachootin and Thomson 1981) was now strongly skewed towards formation of a laterally reduced foot with an strong and elongated central axis. As increased size was selected for, the developmental system inevitably failed in the lateral digits. Once again, the whole pattern was further reinforced by continued selection for fast running, because the new variants that were being produced turned out to be highly efficient for fast running. The result was production of variants showing even greater lateral digit reduction, and the loss of digits two and three (except for the chondrified rudiment of metapodial four). Although the focus of selection had changed from emphasis on proportions to size, the effect on the developmental systems of the limb bud was the same. The result was the one-toed foot, and at this point, even selection for dwarfed body size (in the *Nanhippus* lineage) could not affect the new system of pattern control in the limb bud.

This model can be tested in a number of ways, as follow:

1. It can be tested against a more complete descriptive account of the ontogeny of the normal horse limb.
2. While it seems unlikely that it could, in the near future, be tested experimentally through manipulation of early horse embryo material, this may eventually be possible. However, the study of mutant conditions can significantly assist in the study, particularly if, as is the case with syn in cattle, a true breeding strain of polydactylous horses could be developed.
3. It can be tested comparatively through the fossil record. For example, if the above scheme is correct, then one would expect to find roughly similar patterns of evolution in other organisms where the first mode of selection is selection for change in limb proportion. In fact, the bird wing is an example of strong reduction of lateral elements and fusion of elements in association with proximo-distal elongation of the forelimb. On the other hand, we would expect a significantly different pattern of changes in phenotype to have occurred if the organisms had first been subject to strong selection for size increase and only secondarily to selection for change in limb proportion. Perhaps an example here would be the chalicotheres.

CONCLUSION: PARALLELISM AND CONVERGENCE IN UNGULATE LIMBS

If the preceding discussion is correct, parallelism and convergence in systems like the ungulate limb are strongly predicted. Parallel and convergent evolution are not caused by any of the mechanisms that have been proposed in the past. They are not caused by the chance evolution of exactly the same structural gene mutation(s) independently in separate lineages. They are not caused merely by a coincidence in response by different genomes to a common very powerful directional selection. They do not owe their origin to the fact that there is only one functional solution to a particular adaptive problem. Parts of the cause may be found in these explanations, but what must be added is the fact that the pattern of phenotypic change is the direct consequence of a general set of changes to the morphogenetic pattern control mechanism of the early limb bud. There is only one general phenotypic consequence of a particular set of insults to the developmental system of the limb. The same developmental insults can be applied to a variety of lineages and the result should be roughly the same.

This is not merely an example of developmental constraint in the sense of passive limitation of the range of future changes. It is also a kind of developmental forcing in that there is a feedback: change in proportions—skewed introduction of variation—phenotypic shift—functional shift (spring ligament)—new modes of selection (size)—variation further biased in the same direction, etc.

In this scheme of things, the shape of evolution is caused not just by selection but by the internal properties of developmental systems and the scope of potential response to stress. Because these properties are common to a wide range of organisms, parallelism and convergence are a direct consequence even when the cause of the particular stresses in given lineages are different. It should be noted, however, that the case of evolution of a complex set of adaptations used as an example here may be somewhat easier to explain than others because the phenotypic changes all involve reductions and reorganizations rather than elaborations of complexity. But I believe that the general principle discussed here is a fundamental one: all major divergences in phenotype are caused through the fixation of changes in the early pattern control mechanisms of development rather than late differentiation stages. Lineages diverge through such changes and lineages evolve in parallel or convergently when the common features of early developmental processes cause them to respond to disruption in common ways. Trends (which are only recognizable ex post facto) occur when these commonalities force the response of the developmental system into a single mode even under a succession of selective modes and especially when there is feedback in the system between the direction of selection to the direction of variation.

In the end, approaches like this show us that the traditional dichotomies between internal and external, or microevolution and macroevolution, as applied to mechanisms of evolution are meaningless. They are not extremes or points on a continuum but, rather, they are levels in a hierarchy. No one mechanism is more valid than another because they operate in different realms. A complete theory of evolution will eventually incorporate every level and enunciate the mechanisms, rules, and interconnections of each.

REFERENCES

Alberch P, Gale E (1983): Size dependence during the development of the amphibian foot. Colchicine induced digital loss and reduction. J Embryol Exp Morph 76:177–197.

de Beer GR (1958): "Embryos and Ancestors." New York: Oxford Univ Press.

Camp CL, Smith N (1942): Phylogeny and function of the digital ligaments of the horse. Mem Univ Calif 13:69–124.

Carlens O (1927): Beitrag zur Kenntnis der embryonalen Entwicklung des Extremitatenskeletts beim Pferd und Reind. Gegenbaur's Morphologisches Jahrbuch 58:153–196.

Dobzhansky T (1937): "Genetics and the Origin of Species." New York: Columbia Univ Press.

Ewart JC (1894a): The development of the skeleton of the limbs of the horse, with observations on polydactyly. J Anat Physiol 28:236–256.

Ewart, JC (1894b): The second and fourth digits in the horse: Their development and subsequent degeneration. Proc Roy Soc Edin 20:185–191.

Fisher RA (1930): "Genetical Theory of Natural Selection." ed 2. New York: Dover.

Gruneberg H (1963): "The Pathology of Development. A study of Inherited Skeletal Disorders in Animals." Oxford: Blackwell Scientific Publications.

Gruneberg H, Huston K (1965): The development of bovine syndactylism. J Embryol Exp Morph 19:251–259.

Hermans WA (1969): Een hereditaire anomalie big Shetland Ponies. Vijschr Diergeneesk 94:989–996.
Hinchliffe JR, Johnson DR (1980): "The Development of the Vertebrate Limb." London: Oxford Univ Press.
Huxley JS (1942): "Evolution, the Modern Synthesis." London: Allen and Unwin.
Jepsen GL, Mayr E, Simpson GG (1949): "Genetics, Paleontology and Evolution." Princeton: Princeton Univ Press.
Johnson JL, Leipold HW, Guffy MM, Dennis SM, Schalles RR, Mueller RE (1982): Characterization of bovine polydactyly. Bovine Practice 3:7–114.
Leipold HW, Dennis SM, Huston K (1973): Syndactyly in cattle. Vet Bull 43:399–403.
Lovtrup S (1987): "Darwinism: The Refutation of a Myth." London: Croom, Helms.
MacFadden BJ (1986): Fossil horses from *"Eohippus"* to *Equus:* Scaling, Cope's Law, and the evolution of body size. Paleobiology 12:355–369.
MacFadden BJ (1988): Horses, the fossil record, and evolution. Evolutionary Biology 22: 131–158.
Marsh OC (1892): Recent polydactyle horses. Am J Sci 43:23–355.
Mayr E (1963): "Animal Species and Evolution." Cambridge (MA): Harvard Univ Press.
Muller G (1986): Effects of skeletal change on muscle pattern formation. Bib Anat 29:891–108.
Prentiss CW (1903): Polydactylism in man and the domestic animals, with especial reference to digital variations in swine. Bull Mus Comp Zool Harvard 40(6):245–313.
Prothero DR, Shubin N (1986): Epigenetic control of lateral toe reduction in perissodactyls. Fourth N Amer Paleontol Conv Abstr, A37.
Rachootin SR, Thomson KS (1981): Epigenetics, paleontology and evolution. In Scudder GL, Reveal CL (eds): "Evolution Today." Pittsburgh: Hunt Institute.
Radinsky L (1984): Ontogeny and phylogeny in horse skull evolution. Evolution. 38:1–15.
Shamis LD (1985): Complete ulnas and fibulas in a pony foal. J Am Vet Med Assn 186:802–804.
Shubin N, Alberch P (1986): A morphogenetic approach to the origin and basic organization of the tetrapod limb. Evolutionary Biology 20:319–388.
Simpson GG (1961): "Horses." New York: Doubleday.
Smith JM (1966): "The Theory of Evolution." ed 2. London: Penguin Books.
Smith JM, Burian R, Kauffman S, Alberch P, Campbell J, Goodwin B, Lande R, Raup D, Wolpert L (1985): Developmental constraints and evolution. Q Rev Biol 60:265–287.
Speed JG (1958): A cause of malformation in the limbs of Shetland Ponies with a note on its phylogenetic significance. Brit Vet J 114:18–22.
Thomson KS (1986): On the relationship between development and evolution. Oxford Surveys in Evolutionary Biology. Vol. 2.
Thomson KS (1988a): Fisher's microscope, and the gradualists' dilemma. American Scientist 76:500–502.
Thomson KS (1988b): "Morphogenesis and Evolution." New York: Oxford Univ Press.
Toole B, Jackson G, Gross J (1972): Hyaluronate in morphogenesis: Inhibition of chondrogenesis in vitro. Proc Natl Acad Sci USA 69:1384–1386.
Vrba ES, Eldredge N (1984): Individuals, hierarchies, and processes: Towards a more complete evolutionary theory. Paleobiology 10:146–171.
Waddington CH (1967a): [Discussion.] In Moorehead PS, Kaplan MM (eds): "Mathematical Challenges to the Neo-Darwinian Interpretation of Evolution." Philadelphia: The Wistar Institute of Anatomy and Biology. [Reprinted 1985 by Alan R. Liss, Inc.]
Waddington CH (1967b): "The Strategy of the Genes." London: Allen and Unwin.

New Perspectives on Evolution, pages 123-137

New Perspectives on the Molecular Evolution of Genes and Genomes

DANIEL L. HARTL
Department of Genetics, Washington University School of Medicine, St. Louis, Missouri 63110-1095

INTRODUCTION

Evolutionary biology and molecular biology are becoming synthesized in a way that requires new evolutionary concepts to be developed, new issues to be addressed, and new experimental approaches to be implemented. Two examples of this synthesis concern mechanisms of enzyme evolution and the evolution of genome structure. Recent studies suggest that new proteins often result from the combination of existing functional units rather than from successive single amino acid replacements. For such composite molecules to function, small segments of the protein must fold mainly in accord with local amino acid interactions. In support of this hypothesis, when multiple amino acid replacements were made in several ahelical regions of *E. coli* alkaline phosphatase, new amino acid sequences that form α helices in their native folding context often retained function in the chimeric protein. Relative to genome evolution, the distribution of insertion sequences in the genome of *E. coli* is adequately accounted for by the hypothesis of selfish DNA, in which transposable elements are maintained by replication and transposition in spite of dysgenic consequences. This finding does not imply that beneficial effects of transposable elements do not occur, only that they do not normally occur with sufficient intensity or duration to have a significant effect on the statistical distribution of copy numbers. There is a serious gap in understanding chromosome organization and evolution at the megabase level between light microscopy and DNA sequences. The recent development of a cloning system using yeast artificial chromosome (YAC) vectors, which allows the cloning of exogenous DNA inserts of hundreds of kilobase pairs, may permit the analysis of genome evolution at this megabase level.

A few months ago James F. Crow and William F. Dove invited me to write a piece for the *Perspectives* department of the journal Genetics, which is

Supported by grants GM40322 and GM30201 from the National Institutes of Health.

devoted to short essays with "anecdotal, historical and critical commentaries on Genetics." The occasion was the 15th anniversary of the publication of a paper by Barry G. Hall and myself on enzyme evolution, which has since been developed, largely by Hall, into an elegant and powerful system for experimental evolution (Hartl 1989a). In mulling over the conceptual framework of the essay, I recalled the 1966 Wistar Symposium on *Mathematical Challenges to the Neo-Darwinian Interpretation of Evolution* (Moorhead and Kaplan 1967), because it included a discussion of theories of protein evolution and their shortcomings. No sooner had I ordered the library's copy of the symposium sent up, when Leonard Warren telephoned to invite me to participate in this, the anniversary symposium. This weirdly telepathic history is how I came to be here.

In keeping with the theme *New Perspectives on Evolution,* my assignment was to represent the transition that evolutionary biology has undergone in the past two decades. In describing this transition the earlier Wistar Symposium provides a suitable starting point. Even though the purpose of the 1966 symposium was not to identify the major issues in evolutionary biology—it was rather to confront the major challenges—some of the most influential evolutionary biologists of the time participated, and the great themes were touched upon, at least in the discussions if not in the formal papers.

In summarizing the major themes of the earlier meeting, it becomes pretty clear that there has indeed been a great shift in emphasis in evolutionary biology since 1966, and certainly a shift toward reductionism. In some sense, 1966 marks the climactic end of the evolutionary synthesis begun in the 1930s with the monumental works of Fisher, Haldane and Wright, and the beginning of a new phase in which evolutionary biology began a new synthesis (anastomosis might be a better word) with molecular biology. The watershed was marked by two events: First, the systematic use of protein electrophoresis to estimate the amount of genetic variation present in natural populations (Lewontin and Hubby 1966, Harris 1966); and second, the proposal that many of the genetic differences within and among populations are selectively equivalent (Kimura 1968, King and Jukes 1969). Granted that historical currents are sometimes difficult to spot when you're swimming among them, in this case the sea change in evolutionary biology was apparent: The field was going molecular.

The shift toward molecular biology provided powerful new research tools with which to study one of evolutionary biology's hoariest issues: How much genetic variation exists in natural populations and how is it maintained? Understanding genetic variation dominated population genetics and much of evolutionary biology throughout the decade of the 1970s (Lewontin 1974). However, what is probably more significant in the long run is that the shift toward molecular biology also generated entirely new kinds of concepts and issues that evolutionary biologists needed to ponder and incorporate into their world view.

The principal focus of evolutionary biology today remains much as Waddington summarized it colorfully in the 1966 Wistar Symposium: "The whole real guts of evolution is how do you come to have horses and tigers and things" (Moorhead and Kaplan 1967:14). However, the manner in which this question is addressed has changed dramatically. Here were some of the main themes of the 1966 symposium:

1. Protein evolution. At the time, truly novel enzyme activities were assumed to originate by chance from successive single amino acid substitutions. Eden (1967) pointed out a major difficulty, illustrated by the calculation that a particular sequence of 100 amino acids in a functional polypeptide will occur by chance with a probability of only about 10^{130}. In view of this vanishingly small chance, Eden (1967:7) argued that "either functionally useful proteins are very common . . . so that almost any polypeptide [of random amino acid sequence] . . . has a useful function to perform, or else . . . there exist certain strong regularities for finding useful paths [of protein evolution]." Wright (1967) emphasized that polypeptides are not assembled at random but selected step by step from pre-existing ones, and he likened natural selection to the game of Twenty Questions, in which it is possible to arrive at the correct sequence of 100 amino acids in a polypeptide by a series of 500 questions answered yes or no. Wright's point is valid, but many novel enzyme functions may require multiple amino acid substitutions and therefore not be selectable step by step.
2. Population structure. This was a key issue in population genetics since the Wright-Fisher controversy of the 1930s (Provine 1986). It includes effective population number, pattern of subdivision into subpopulations, and amount and pattern of migration among subpopulations.
3. Levels of selection. In the 1960s Darwinian selection meant selection based upon each individual's own phenotypic merit, and ideas about selection possibly occurring at other levels were still rather vague. Inchoate in some of the discussions, for example Wald (1967) on vicarious selection, were concepts of group and kin selection. In the 1960s there was no consensus that most genomes contain DNA sequences that are selfish in the sense of maintaining themselves by replication and transposition in spite of dysgenic consequences.
4. Genetic variation. Most studies of genetic variation have focused on the types, amounts and maintenance of genetic variation within natural populations. While genetic variation was not emphasized in the 1966 meeting, it underlay much of the discussion. Stimulated by the application of protein electrophoresis to natural populations (Lewontin and Hubby 1966, Harris 1966) and the challenge to orthodox Darwinism presented by the neutral theory (Kimura 1968, King and Jukes 1969),

understanding the amount and causes of genetic variation was to become a major preoccupation of population genetics in the 1970s.

5. The theory of gene frequencies. Understanding processes at the level of gene frequencies has been a major objective of population genetics since the seminal works of Wright, Fisher and Haldane (see Provine 1971). At the Wistar Symposium, Lewontin (1967) used computer simulation to study the consequences when selection coefficients vary in space and time, and he showed that the outcome of selection can depend on the historical pattern of selection coefficients. At the time there was considerable excitement over the potential of high speed computers in the solution of otherwise intractable problems in theoretical population genetics.
6. Speciation and macroevolution, or how you come to have horses and tigers and things. Traditional evolutionary biology has always emphasized the formation of new species and higher taxonomic categories. Waddington (in Moorhead and Kaplan 1967:75) made it clear that understanding evolution at this level requires understanding mechanisms of development. Progress has been tremendous during the past 20 years, but there is still a long way to go in uniting developmental approaches with evolution.
7. Origin of life. This borderline field between biophysical chemistry and evolutionary biology has undergone major advances since the 1966 symposium. Chief among these has been the discovery that RNA can act as an enzyme as well as an informational macromolecule and so provide a plausible candidate for the original genetic material (Cech 1987).
8. Epistemological issues in evolutionary theory. Is the theory of natural selection fundamentally tautological? If not, does the neo-Darwinian theory provide an adequate understanding of adaptive processes?

The themes in this list are not mutually exclusive, nor do they comprehend everything that was discussed. They do, however, provide an overview of the kinds of problems that were uppermost in the minds of leading evolutionary biologists more than 20 years ago. I regard it as favorable testimony to the health and vigor of evolutionary biology that real progress has been made in defining and understanding each of these issues since the Wistar Symposium of 1966. Much of this progress in population genetics has been summarized in my recent textbooks (Hartl 1988, Hartl and Clark 1989).

I define new perspectives in evolutionary biology as resulting from new concepts that have been incorporated into evolutionary thinking or from new experimental techniques that have opened up promising directions. I am going to focus on two specific examples of work from my own laboratory. The reasons are familiarity, convenience, and a desire to be concrete. The examples typify some of the exciting new directions in which many evolutionary biologists are presently engaged in a variety of systems. My examples concern mechanisms of enzyme evolution and the evolution of genome structure.

Mechanisms of enzyme evolution. There are two distinct aspects of enzyme evolution: How novel enzyme functions arise in the first place; and, having once arisen, how they are progressively refined. The classical view is that these are two different aspects of the same fundamental mechanism of successive random amino acid replacement, with a lucky few replacements improving the function of existing enzymes and still fewer creating useful enzyme functions de novo. It is the apparent unlikelihood of creating useful new functions by random amino acid replacements that prompted Eden's (1967) challenge to Darwinism, and his dilemma still appears valid for the creation of functions requiring more than a few simultaneous amino acid replacements in order to occur. In the widely accepted theory of enzyme evolution by means of duplication and divergence, new catalytic activities are supposed to evolve by random amino acid replacements in duplicate copies of preexisting genes. However, Eden's dilemma suggests that any new functions evolved by this mechanism are likely to be no more than a few mutational steps away from the starting point. Of course, the dilemma is circumvented when new enzyme functions arise by chance during the course of evolution for the progressive improvement of enzyme functions that already exist.

Recent studies of the structure and nucleotide sequence of enzyme-coding genes, also supported by crystallographic studies of three-dimensional structure, suggest that novel protein functions may often arise by the combination of existing functional units from other proteins rather than arising *de novo* from random amino acid replacements in a single protein. From the standpoint of enzyme structure, domains with similar folding characteristics and similar functions are often found in otherwise unrelated proteins (Phillips et al. 1983). From the standpoint of gene structure, the exons present in many eukaryotic genes are often correlated with folding domains in the mature protein, which suggests that the genes were originally assembled piecemeal from smaller units capable of relatively autonomous folding and function (Gilbert 1978). A remarkable example supporting the genes-in-pieces model of protein evolution is the low-density lipoprotein receptor gene, which contains exons that are clearly paralogous (homologous because of gene duplication) with exons in genes for components of complement, bloodclotting factors, and epidermal growth factor (Sudhof et al. 1985).

While comparative data can provide support for the genes-in-pieces model, it is difficult to define a direct experimental test. A more directly testable hypothesis has recently been suggested by Brenner (1988). He proposes a principle of local functionality, in which the folding of small segments of polypeptide is determined mainly by local interactions within each segment. Local functionality allows discrete regions of a polypeptide to fold and function with relative autonomy in composite molecules. The joining of such local functional units early in evolution provided the physical basis for assembling larger functional domains, which have been preserved, in many cases, in the pattern of exons and introns found in present-day genes.

One aspect of local functionality that can be tested is the implication that certain small segments of functionality may be replaced with segments of comparable local structure from totally unrelated molecules, without completely destroying function. The multiple adjacent amino acid replacements needed to test the hypothesis would be difficult to create by purely genetic means, but they can be created by the procedure of oligonucleotide site-directed mutagenesis. On almost any model of protein evolution, multiple amino acid replacements that do not preserve local structure would be expected to destroy function. If replacements that preserve local structure also destroy function, then the principle of local functionality is cast into doubt.

As an experimental system for testing local functionality, we have chosen the enzyme alkaline phosphatase in *Escherichia coli* and have focused on several regions of the molecule that form α helices (DuBose and Hartl 1989). The rationale for this choice is that the experimental manipulations are straightforward to carry out in *E. coli,* the enzyme is relatively easy to purify, and its three-dimensional structure is well documented (Sowadski et al. 1985). Regions of α helix seemed a favorable choice for initial experiments because α-helical secondary structure are very well defined although their degree of folding autonomy within larger protein units has not been evaluated.

Using oligonucleotide site-directed mutagenesis, multiple amino acid replacements were created for each of three α-helical segments in alkaline phosphatase, each seven amino acids in length. Details of the experiments and the results can be found in DuBose and Hartl (1989). For present purposes, the salient results are that, while replacements with random amino acid sequences always eliminated enzyme function, replacements with unrelated amino acid sequences known to form α-helices in their native folding context often preserved function in the chimeric protein. Two of the sequences used in the replacements were taken from other α-helical regions in alkaline phosphatase; one was from a helical segment in the unrelated protein lysozyme from bacteriophage T4 (Fig. 1), and the other was a synthetic peptide known to form a helical structure in solution (Regan and DeGrado 1988). Interestingly, while the native α-helical structures could be used in the replacements without loss of function, the artificial helix eliminated function completely.

We regard these results as strongly supporting the principle of local functionality, at least with regard to helical segments. Although some of the helical replacements reduced enzyme activity substantially, the chimeric enzymes retained enough activity that they could easily provide a starting point for further improvement in activity by means of natural selection acting on conventional single amino acid replacements.

Since our experimental test used a specific protein and a specific subset of structural units that were manipulated, the evidence for the principle of local functionality should be generalized only with caution. However, as experience with different types of structural units within different types of proteins ac-

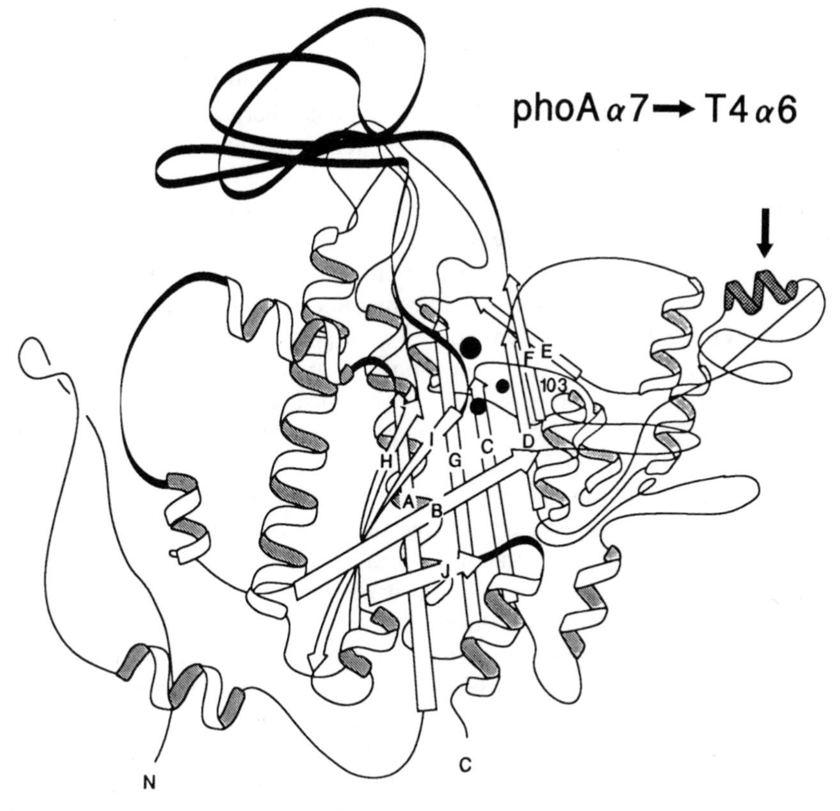

Fig. 1. Two-dimensional projection of monomeric unit of alkaline phosphatase. The enzyme still functions when helix 7 **(arrow)** is replaced with an unrelated helix derived from bacteriophage T4 lysozyme. How much latitude is allowable in the replacement of such structural units remains to be determined.

cumulates, patterns may begin to emerge that will help to refine the model and to identify more precisely the critical structural parameters.

Although the genes-in-pieces model for the evolution of novel functions fills an important gap in evolutionary theory, this does not diminish the critical role that single amino acid replacements play in refining new functions after they occur. The critical role of individual amino acid replacements in experimental enzyme evolution is well demonstrated by the evolved β-galactosidase coded by the ebgA gene in E. coli. This gene codes for a cryptic β-galactosidase that has little activity toward lactose as a substrate, but under suitable conditions mutations can be selected that permit growth on lactose as the sole carbon source (Hartl 1989a). When strains that synthesize the wildtype *ebgA* enzyme constitutively are selected for growth on lactose, two types of mutations result.

The first gives good growth on lactose but not lactulose (galactosyl-fructose), the second gives moderate growth on lactose but good growth on lactulose (Hall 1981). These enzymes differ by single nucleotide substitutions at opposite ends of the *ebgA* gene (Hall and Zuzel 1980). When both of the mutations are brought together in the same gene, they confer a new substrate specificity for lactobionic acid. That is, the single amino acid replacements improve preexisting specificities for lactose and lactulose, respectively, but the double mutant does something entirely new.

At the amino acid level, the enzyme that allows growth on lactose but not lactulose contains a single amino acid replacement of asparagine for aspartic acid at position 92 in the 1032 amino acid *ebgA* polypeptide (Stokes, Betts, and Hall 1985; Hall, personal communication). In the enzyme that allows growth on both lactose and lactulose, there is a replacement of cysteine for tryptophan at position 977, and some isolates also contain a serine to glycine replacement at position 978 (Hall, personal communication). Interestingly, these replacements are not located in proximity to the active site of the enzyme.

From an evolutionary standpoint, the functional effects of amino acid replacements that are polymorphic in natural populations are of great interest. In the polarized terms of the selection-neutrality controversy of the 1970s, these polymorphisms might be the raw material for adaptive evolution or be adaptively neutral molecular background noise. This issue has been very difficult to resolve, in part because selective effects that are important may have magnitudes well below the limit of experimental detection.

Fortunately, there is a statistical test that can estimate selection coefficients much smaller than are detectable experimentally (Sawyer et al. 1987b). The test uses nucleotide sequence data from naturally occurring alleles, and it compares the sampling configurations of silent polymorphisms with those of amino acid polymorphisms in the same gene. The sampling configuration of a nucleotide site relates to the number of different nucleotides at the site that occur in the sample and the number of individuals with each nucleotide. The sampling configurations of the silent polymorphisms (those coding for identical amino acids) can be tested against the theoretical expectations of near neutrality and used to estimate parameters of mutation and random genetic drift. A sampling distribution with these parameters is then compared with the observed sampling configurations of the amino acid polymorphisms, and the model modified if necessary by the incorporation of a selection parameter.

This test of selection is intuitively appealing in that it compares amino acid polymorphisms with silent polymorphisms in the same gene. Aspects of population structure that affect the silent polymorphisms are expected to affect amino acid polymorphisms in the same way, and silent polymorphisms, if not completely neutral, are nevertheless expected to be less strongly selected than amino acid polymorphisms.

In the analysis of Sawyer et al. (1987b), 768 homologous nucleotide positions were compared among seven naturally occurring alleles *gnd,* which codes for 6-phosphogluconate dehydrogenase, in *E. coli.* These sequences included 12 amino acid polymorphisms and 78 silent nucleotide polymorphisms. Among the silent polymorphisms, 34/66 occurred in a "6,1" sampling configuration, in which six alleles exhibited the majority nucleotide at the site and one allele exhibited a different nucleotide. Among the amino acid polymorphisms, 12/12 had the "6,1" configuration. Comparing 34/66 against 12/12 using Fisher's exact test gives a P value of 0.0026, so clearly the sampling distributions of silent sites and amino acid polymorphisms are incompatible. Various refinements of the test also support this conclusion (Sawyer et al. 1987b).

The discrepancy is in the direction of too few alleles exhibiting the minority amino acid at polymorphic sites. This is consistent with a selection model in that at least some of the amino acid replacements are mildly deleterious. In the *gnd* data, no more than about half of the amino acid polymorphisms can be imputed selective effects as small as those affecting silent polymorphisms. Alternatively, if they are all mildly deleterious, the sampling configurations are consistent with an average selection coefficient of 1.6×10^{-7} (Sawyer et al. 1987b). A similar analysis of amino acid polymorphisms in the *phoA* gene of *E. coli* suggests deleterious selection of a comparable magnitude, 5.5×10^{-7} (Sawyer, DuBose and Hartl, unpublished).

Our statistical analyses of protein polymorphisms are completely consistent with an overview recently articulated by Crow (1989): "I think the best current interpretation is that most of the observed protein polymorphism is a mixture of nearly neutral alleles, whose frequency is determined by mutation and random drift, and rare deleterious alleles maintained by mutation-selection balance." However, I would go a step further and claim that many of the nearly neutral protein polymorphisms have the potential to undergo selection in a changed environment. This assertion is based on studies of the selective effects of protein polymorphisms in chemostats. In the experimental system, isogenic pairs of *E. coli* strains differing in an enzyme polymorphism were made to compete for substrates that utilize the enzyme in their metabolism. The main finding was that naturally occurring enzyme variants usually produced effects on fitness that were undetectable under the conditions usually considered optimal for *E. coli,* but many of the enzyme variants had significant effects on fitness when the conditions were altered, for example, when competition was for an unusual substrate (Dykhuizen and Hartl 1980, Hartl and Dykhuizen 1981). The result implies that naturally occurring genetic variants, many of which may be selectively neutral or nearly neutral under the prevailing mosaic of environments, may nevertheless have a latent potential for selection that can be expressed under alternative environmental conditions (Hartl and Dykhuizen 1984, 1985). The implications of this principle for general evolution have been discussed by Stebbins and Hartl (1988) and Kimura (1989).

In a broader context, the finding that many naturally occurring enzyme variants have very small effects on fitness is a consequence of the functional interactions among enzymes in complex metabolic pathways and their partitioning of the control of flux. The quantitative analysis of metabolic interactions is the province of metabolic control theory (Kacser and Porteous 1987), which has been discussed in an evolutionary context by Hartl et al. (1985), Dean et al. (1988) and Hartl (1989a, 1989b).

Mechanisms of genome evolution. Another area for the fruitful merging of evolutionary and molecular biology concerns processes of genome evolution. For example, the finding that transposable elements are widespread in virtually all organisms was a stimulus for research into the mechanisms by which transposable elements are maintained in populations and the factors that determine their abundance and spatial distribution within the genome (reviewed in Ajioka and Hartl 1989). More recently, the development of yeast artificial chromosomes for cloning DNA molecules of hundreds of kilobases (Burke et al. 1987, Garza et al. 1989) opens up the prospect of studying the evolution of genome organization at the critical megabase level between the low resolution of light microscopy and the high resolution of nucleotide sequences.

In regard to transposable elements, a provocative hypothesis is that they do not, on the whole, increase the fitness of the organisms in which they occur, but are rather selfish elements in the genome maintained by replication and transposition (Doolittle and Sapienza 1980, Orgel and Crick 1980). We have evaluated this hypothesis using extensive data on the distribution of insertion sequences in the genome of *E. coli.* These are 13 kilobase transposable elements terminating in short inverted repeat sequences and usually containing one or more coding regions for proteins needed for transposition and, in some cases, the regulation of transposition (Galas 1989). The primary forces determining the population dynamics of insertion sequences are transposition, regulation of transposition, horizontal dissemination among hosts by means of transmissible plasmids containing the elements, and any effects the elements may have on fitness.

Natural isolates of *E. coli* are polymorphic for the presence or absence of insertion sequences. For example, in studies of the six insertion sequences IS*1,* IS*2,* IS*3,* IS*4,* IS*5* and IS*30* among 72 natural isolates comprising the ECOR reference strains (Ochman and Selander 1984), while the strains exhibited 1055 occupied chromosomal sites and 118 occupied plasmid sites, the numbers of strains lacking copies of the individual insertion sequences were 11, 28, 23, 43, 46 and 36, respectively (Sawyer et al. 1987a). This immediately suggests that most copies of insertion sequences are not beneficial to fitness, otherwise most isolates would be expected to contain one or more copies.

In the quantitative analysis of the IS distributions carried out by Sawyer et al. (1987a), the observed distributions were compared with population models in which the number of copies of each IS sequence in a strain changes according to a continuous time branching process with rates of transposition and fitness given by simple functions of the number of copies in the strain. Using this analysis, the insertion sequences could be classified according to the apparent strength of regulation of transposition. By regulation of transposition I mean the tendency for the rate of transposition per element to decrease as total copy number increases. The elements IS*1* and IS*5* fit population models with weak or no regulation of transposition; IS*2,* IS*4* and IS*30* fit models with moderate regulation; and IS*3* and IS*150* (also known as IS*103,* Hall et al. 1989) fit a model with strong regulation (Fig. 2).

In terms of effects on fitness needed to account for the observed numbers of elements, all models required a small decrease in fitness among infected strains, which is consistent with the selfish DNA hypothesis. As a function of copy number, the best models for IS*1* or IS*5* required that fitness decrease as a linear or quadratic function of copy number, whereas the best models for the other insertion sequences required that fitness decrease modestly, if at all, with

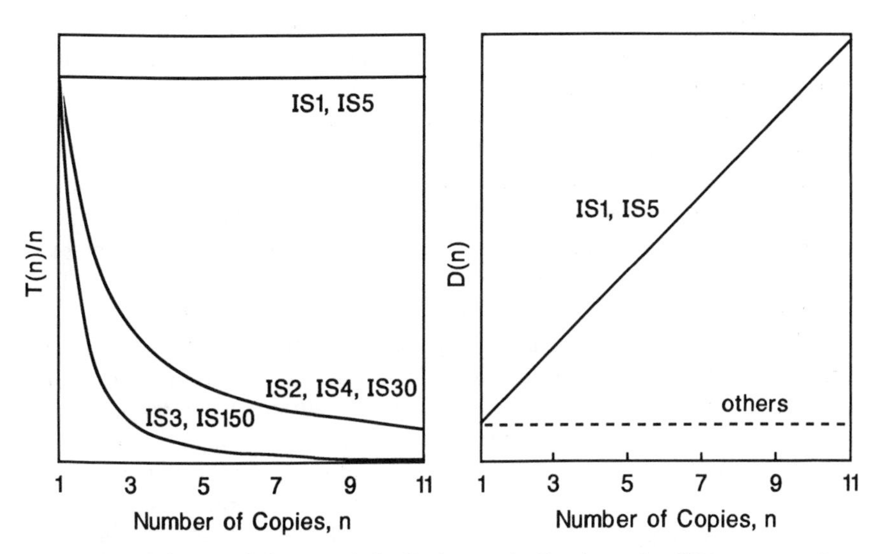

Fig. 2. Best-fitting population models for IS elements in *E. coli* require different assumptions about regulation of transposition **(left)** and effects on fitness **(right).** The models for IS*1* and IS*5* have essentially no trans-regulation of transposition, as the transposition rate per element $(T(n)/n)$ is approximately constant. IS*2,* IS*4* and IS*30* have moderate regulation ($T(n)/n$ decreases about as $1/n$), and IS*3* and IS*150* have strong regulation ($T(n)/n$ decreases about as $1/n^2$). The inferred differential death rate $D(n)$ increases linearly with copy number for IS*1* and IS*5* but is essentially independent of copy number for the other IS elements studied.

increasing copy number (see Fig. 2 and Sawyer et al. 1987a, Ajioka and Hartl 1989).

The statistical analysis also revealed an unexpected correlation among the ECOR strains, namely that unrelated insertion sequences tend to occur together (Hartl and Sawyer 1988). That is, strains infected with one insertion sequence are more likely to be infected with a different one. Averaged across strains, the mean pairwise correlation coefficient in the occurrence of insertion sequences was nearly 20 percent. On the surface very mysterious, this correlation proved to be a necessary consequence of the mode of dissemination of insertion sequences among strains by means of transmissible plasmids, and the observed frequencies of insertion sequences in plasmids can account quantitatively for the correlations (Hartl and Sawyer 1988).

In summary, models of selfish DNA are completely adequate to explain quantitatively the observed numbers of insertion sequences among natural isolates of *E. coli.* The analyses cannot exclude occasional beneficial effects on fitness resulting from lucky insertion sites or from the mobilization of advantageous genes, for example, antibiotic resistance genes. Such effects have been very well documented in a variety of situations (see papers in Berg and Howe 1989). However, invoking beneficial effects is not necessary to explain the observed copy numbers. I conclude that, while beneficial effects do occasionally occur, and may in fact be very important in the evolution of transposable elements, they do not normally occur with sufficient intensity or duration to have a significant effect on the distribution of copy numbers among individuals. These issues are discussed in greater detail in Ajioka and Hartl (1989).

Among the aspects of genome evolution that have not yielded to conventional molecular biology are those involving the organization of large tracts of DNA, for example, a and b heterochromatin, centromere regions, telomere regions, or any level of structure encompassing more than a few hundred kilobases of DNA. In some cases the limitation results from genetic instability, or other sources of difficulty, in cloning the tracts, but in other cases the tracts are simply too lengthy to be isolated and manipulated in conventional cloning systems. Paradoxically, for studying genome evolution at this level, conventional methods are either too powerful or not powerful enough. Light microscopy is appropriate at the level of visible chromosomes, conventional cloning at the level of short-range restriction maps or nucleotide sequences. In between is a level consisting of a few hundred to a few thousand kilobases that is not easily accessible with either method.

The recent development of a cloning system using yeast artificial chromosome (YAC) vectors (Burke et al. 1987) encourages some optimism that the megabase gap in analytical methods can be bridged. YAC vectors can accept exogenous DNA inserts of hundreds of kilobase pairs, and at least some genomic sequences that are difficult to clone in other vectors can be cloned in YACs (Garza et al. 1989).

Although the YAC system has generated immediate excitement because of its application to the physical analysis of complex genomes, such as nematodes, *Drosophila,* the mouse, and humans, I believe that one of the eventual uses of the method will be in the study of genome evolution at the megabase level. I also believe that *Drosophila* will be an ideal organism for such studies because of its polytene salivary gland chromosomes. Since the approximately 5000 bands in the salivary chromosomes each contain an average of about 20 kilobases, YACs containing a few hundred kilobases would be expected to cover 10-20 bands and therefore provide a contiguous, cytologically visible unit for comparison among different species.

The feasibility of this approach has been demonstrated using a YAC library containing DNA from *Drosophila melanogaster* (Garza et al. 1989). In situ hybridizations of approximately 60 YAC clones averaging about 200 kilobase pairs have given many examples of YACs containing multiple salivary bands. Collectively, these clones include about 10 percent of the euchromatic genome, and as overlapping clones are found in the ongoing studies, they will provide a framework for the study of genome organization and evolution at the megabase level. This is an opportunity in molecular evolutionary biology that was scarcely conceivable at the time of the Wistar Symposium in 1966.

REFERENCES

Ajioka JW, Hartl DL (1989): Population dynamics of transposable elements. In Berg DE, Howe MM (eds): "Mobile DNA." Washington, DC: Amer Soc Microbiol.

Berg DE, Howe MM (eds) (1989): "Mobile DNA." Washington, DC: Amer Soc Microbiol.

Brenner S (1988): A tale of two serines. Nature 334:528–530.

Burke DT, Carle GF, Olson MV (1987): Cloning of large segments of exogenous DNA into yeast by means of artificial chromosome vectors. Science 236:806–812.

Cech T (1987): The chemistry of self-splicing RNA and RNA enzymes. Science 236:1532–1539.

Crow JF (1989): Twenty-five years ago in Genetics: The infinite allele model. Genetics 112:631–634.

Dean AM, Dykhuizen DE, Hartl DL (1988): Theories of metabolic control in quantitative genetics. In Weir BS, Eisen EJ, Goodman MM, Nankoong G (eds): "Proc 2nd Int Conf Quantitative Genetics." Sunderland, MA: Sinauer Associates, pp 536–548.

Doolittle FW, Sapienza C (1980): Selfish DNA: The phenotype paradigm and genome evolution. Nature 284:601–103.

DuBose RF, Hartl DL (1989): An experimental approach to testing modular evolution: Directed replacement of α-helices in a bacterial protein. Proc Natl Acad Sci USA 86: 9966–9970.

Dykhuizen D, Hartl DL (1980): Selective neutrality of 6PGD allozymes in *E. coli* and the effects of genetic background. Genetics 96:801–817.

Eden M (1967): Inadequacies of neo-Darwinian evolution as a scientific theory. In Moorehead PS, Kaplan MM (eds): "Mathematical Challenges to the Neo-Darwinian Interpretation of Evolution." Philadelphia: Wistar Institute Press, pp 5–19. [Reprinted (1985) by Alan R. Liss, Inc., New York.]

Galas DJ (1989): Review of IS sequences. In Berg DE, Howe MM (eds): "Mobile DNA." Washington, DC: Amer Soc Microbiol.

Garza D, Ajioka JW, Burke DT, Hartl DL (1989): Mapping the *Drosophila* genome with yeast artificial chromosomes. Science 246(4930):641–646.

Gilbert W (1978): Why genes in pieces? Nature 271:501.

Hall BG (1981): Changes in the substrate specificities of an enzyme during directed evolution of new functions. Biochemistry 20:4042–4049.

Hall BG, Parker LL, Betts PW, DuBose RF, Sawyer SA, Hartl DL (1989): IS103 a new insertion element in *E coli*: Characterization and distribution in natural populations. Genetics 121:423–431.

Hall BG, Zuzel T (1980): Evolution of a new enzymatic function by recombination within a gene. Proc Natl Acad Sci USA 77:3529–3533.

Harris H (1966): Enzyme polymorphisms in man. Proc Soc Lond [Biol] 164:298–310.

Hartl DL (1988): "Primer of Population Genetics." ed 2. Sunderland, MA: Sinauer Associates.

Hartl DL (1989a): Evolving theories of enzyme evolution. Genetics 122:1–6.

Hartl DL (1989b): The physiology of weak selection. Genome 31:183–189.

Hartl DL, Clark AG (1989): "Principles of Population Genetics" ed 2. Sunderland, MA: Sinauer Associates.

Hartl DL, Dykhuizen DE (1981): Potential for selection among nearly neutral allozymes of 6-phosphogluconate dehydrogenase in *Escherichia coli*. Proc Natl Acad Sci USA 78:6344–6348.

Hartl DL, Dykhuizen DE (1984): The population genetics of *Escherichia coli*. Annu Rev Genet 18:31–68.

Hartl DL, Dykhuizen DE (1985): The neutral theory and the molecular basis of preadaptation. In Ohta T, Aoki K (eds): "Population Genetics and Molecular Evolution." Tokyo: Japan Scientific Societies Press.

Hartl DL, Dykhuizen DE, Dean AM (1985): Limits of adaptation: The evolution of selective neutrality. Genetics 111:655–674.

Hartl DL, Sawyer SA (1988): Why do unrelated insertion sequences occur together in the genome of *Escherichia coli*? Genetics 118:537–541.

Kacser H, Porteous JW (1987): Control of metabolism: What do we have to measure? Trends Biochem Sci 12:5–14.

Kimura M (1968): Evolutionary rate at the molecular level. Nature 217:624–626.

Kimura M (1989): The present status of the neutral theory. In Tokahata N, Crow JF, "Population Biology of Genes and Molecules." Cambridge: Cambridge Univ Press.

King JL, Jukes TH (1969): Non-Darwinian evolution: Random fixation of selectively neutral mutations. Science 164:788–798.

Lewontin RC (1967): The principle of historicity in evolution. In Moorhead PS , Kaplan MM (eds): "Mathematical Challenges to the Neo-Darwinian Interpretation of Evolution" Philadelphia: Wistar Institute Press, pp 81–88. [Reprinted (1985) by Alan R. Liss, Inc., New York.]

Lewontin RC (1974): "The Genetic Basis of Evolutionary Change." New York: Columbia Univ Press.

Lewontin RC, Hubby JL (1966): A molecular approach to the study of genic heterozygosity in natural populations. II. Amount of variation and degree of heterozygosity in natural populations of *Drosophila pseudoobscura*. Genetics 54:595–609.

Moorhead PS, Kaplan MM (eds) (1967): "Mathematical Challenges to the Neo-Darwinian Interpretation of Evolution." Philadelphia: Wistar Institute Press. [Reprinted (1985) by Alan R. Liss, Inc., New York.]

Ochman H, Selander RK (1984): Standard reference strains of *Escherichia coli* from natural populations. J Bacteriol 157:690–693.

Orgel LE, Crick FHC (1980): Selfish DNA: The ultimate parasite. Nature 284:604–607.

Phillips DCM, Sternberg JE, Sutton BJ (1983): Intimations of evolution from the three-dimensional structures of proteins. In Bendall DS (ed): "Evolution from Molecules to Men." Cambridge: Cambridge Univ Press, pp 145–173.

Provine WB (1971): "The Origins of Theoretical Population Genetics." Chicago: Univ Chicago Press.

Provine WB (1986). "Sewall Wright and Evolutionary Biology." Chicago: Univ Chicago Press.

Regan L, DeGrado W (1988): Characterization of a helical protein designed from first principles. Science 241:976–978.

Sawyer SA, Dykhuizen DE, DuBose RF, Green L, Mutangadura-Mhlanga T, Wolczyk DF, Hartl DL (1987a): Distribution and abundance of insertion sequences among natural isolates of *Escherichia coli*. Genetics 115:51–63.

Sawyer SA, Dykhuizen DE, Hartl DL (1987b): A confidence interval for the number of selectively neutral amino acid polymorphisms. Proc Natl Acad Sci USA 84:6225–6228.

Sowadski JM, Handschumacher MD, Murthy HMK, Foster BA, Wyckoff BA (1985): Refined structure of alkaline phosphatase from *Escherichia coli* at 28 A resolution. J Mol Biol 186:417–433.

Stebbins GL, Hartl DL (1988): Comparative evolution: Latent potentials for anagenic advance. Proc Natl Acad Sci USA 85:5141–5145.

Stokes HW, Betts PW, Hall BG (1985): Sequence of the *ebgA* gene of *Escherichia coli*: Comparison with the *lacZ* gene. Mol Biol Evol 2:469–477.

Sudhof TC, Goldstein JL, Brown MS, Russell DW (1985): The *LDL* receptor gene: A mosaic of exons shared with different proteins. Science 228:815–822.

Wald G (1967): The problems of vicarious selection. In Moorhead PS, Kaplan, MM (eds): "Mathematical Challenges to the Neo-Darwinian Interpretation of Evolution." Philadelphia: Wistar Institute Press, pp 59–62. [Reprinted (1985) by Alan R. Liss, Inc., New York.]

Wright S (1967): Comments on the preliminary working papers of Eden and Waddington. In Moorhead PS, Kaplan, MM (eds): "Mathematical Challenges to the Neo-Darwinian Interpretation of Evolution." Philadelphia: Wistar Institute Press, pp 117–120. [Reprinted (1985) by Alan R. Liss, Inc., New York.]

New Perspectives on Evolution, pages 139-154

Evolution of Transposable Elements in *Drosophila*

MARGARET G. KIDWELL
KENNETH R. PETERSON
Department of Ecology and Evolutionary Biology, University of Arizona, Tucson, Arizona 85721

INTRODUCTION

One of the most significant recent developments in the field of genetics has been the change in our concept of the genomes of living organisms, from the essentially static structures that were previously suggested by classical genetic analysis, to ones that are unexpectedly dynamic and fluid. This new understanding has been one of the most interesting features to have emerged from the application of recombinant DNA techniques to genetic analysis. A major factor in this conceptual change has been the discovery of specific genetic determinants that are mobile in the cellular genome (Shapiro 1983). These mobile, or transposable, elements are sequences of varying size and complexity that are able to move into new positions (insertions) or become removed from old positions (excisions) in the genomes of cellular organisms.

Transposable elements were discovered by Barbara McClintock about forty years ago, but her work was first greeted with considerable skepticism and it has only been in the last two decades that there has been a general realization that transposons are widespread in both prokaryotes and eukaryotes. The study of transposable elements is therefore a very young field of science and most of our current knowledge has been accumulated since the last Wistar Symposium. New observations and insights are occurring at such a rapid rate that reviews become outdated in a period of only a few years. Because of the number and diversity of systems, after discussing some general properties of transposable elements in this paper, we will restrict our focus to the evolution of transposable elements in *Drosophila*.

Drosophila transposable elements, especially those in *D. melanogaster*, are among the best studied among all eukaryotic species for a number of reasons.

Supported by U.S. Public Health Service Grant GM-36715.

Element families are numerous and vary widely in structural types. Many of them produce phenotypic manifestations that have consequences for both population and evolutionary biology. The giant salivary gland chromosomes facilitate the study of insertion sites by means of the technique of in situ hybridization to polytene chromosomes.

GENERAL STRUCTURAL PROPERTIES OF TRANSPOSABLE ELEMENTS

Several structural features are shared by most transposable elements. A transposable element is a defined DNA sequence which commonly, but not always, has inverted terminal repeats of up to 40 base pairs, i.e., the sequences at each end are repeated in inverted orientation. The inverted repeats flank a central region containing transposition genes called transposases and, sometimes, additional genetic determinants; characteristically, a duplication of a few base pairs of target (host) DNA is found at the sites of insertion of these elements. These appear as direct repeats that flank the inserted sequence.

Transposable elements are divided into families, the members of which have extensive sequence homology but may differ in size and possibly function. They can vary in size from a few hundred to a few thousand base pairs. These elements are typically present at multiple sites in the genome and the location of these sites often differs from one individual to another within a species. The copy number of most transposable element families does not increase without restraint and a number of mechanisms exist for maintaining copy number within fairly narrow ranges about a mean which varies for each family. Mechanisms of transposition and excision can also vary widely between different families.

TRANSPOSON-MEDIATED EFFECTS ON HOST DNA

Transposable elements can alter both the organization and expression of genes at frequencies that can exceed those of mutation events due to other causes, both spontaneous and induced (e.g., Sankaranrayanan 1988). The mutations and rearrangements mediated by transposable elements usually involve larger segments of DNA than the single or few base pairs that are involved in point mutations. Transposon-mediated mutations therefore have the potential for producing both major disruptions of a functional genome and quantum jumps in an otherwise slow and continuous evolution of DNA sequences. Transposable elements have a rare property essential for successful mutator elements (Leigh 1973). They have complete linkage between the mutator and its induced effects because the act of insertion into a new site may directly produce new mutant phenotypic traits.

A number of DNA structural changes characteristically result from transposon-mediated rearrangements in both prokaryotes and eukaryotes including insertions, excisions, deletions, inversions, duplications and translocations. Insertions of transposable elements can occur into both the coding and regulatory regions of genes. A new insertion into the coding region of a gene frequently leads to inactivation of that gene. Insertions into regulatory regions of genes are likely to be less drastic and can lead to changes of gene expression including the new expression of cryptic genes. The effects of insertions can be reversed by precise excisions which result in the restoration of the wild type genotype and phenotype. Excisions can also be of the imprecise variety which may leave remnants of some part of the original insertion in the host chromosome, or also take out some part of the host sequences, in addition to all or part of the original insertion. Imprecise excisions can lead to a number of phenotypic effects depending to the extent and location of the sequences involved.

Deletions are a common type of mutation associated with resident transposable elements and they characteristically extend from the terminus of the resident element for varying distances into the host DNA. Two elements inserted in opposite orientation in host chromosomes can lead to inversion of parts of each of the elements and the intervening host DNA. The potential for duplication of host sequences exists if two elements flanking the region of interest lie in direct orientation. The regions of homology in the two elements promote unequal crossing over, leading to tandem duplication of intervening sequences. The concerted transposition of a composite transposon, made up of two transposable elements flanking a sequence of host DNA, can result in the translocation of both transposable elements and the intervening sequence to a new location in the host genome.

In some eukaryotes dramatic phenotypic changes associated with transposable element mobilization are induced for which the molecular basis is not yet clear. Examples are the gonadal dysgenic sterility associated with transposition of the *P* element and the F_2 early embryonic lethality associated with transposition of the I element in *D. melanogaster.* It has been speculated that these types of sterility are caused by massive fragmentation of germline chromosomes mediated by these transposable elements at specific developmental stages (Bregliano and Kidwell 1983).

At the level of the organismal phenotype, the mutations and other genomic changes associated with active mobile sequences may be deleterious, advantageous or neutral but, like mutations in general, the majority of changes caused by mobile elements are expected to be deleterious (e.g., Fitzpatrick and Sved 1986). However, a small minority of such changes may be advantageous to the host organism. For example, it has been shown that cells of *Escherichia coli* harboring the transposons *Tn5* or *Tn10* have a growth advantage in chemostat competition over strains that lack these element but are otherwise isogenic (Chao et al. 1983).

THE ORIGIN AND COURSE OF EVOLUTION OF MOBILE SEQUENCES

The origin of transposable elements is currently completely unknown. They must either have evolved from preexisting host genomic sequences or have been introduced from outside the host genome, for example by viral infection. A number of ideas have been proposed for the origin and maintenance of transposable elements that are not necessarily mutually exclusive. It seems that like other living entities, transposable elements may have characteristic life histories that includes a number of discernable evolutionary stages, some of which may last for millions of years. Their phylogenies may track the phylogenies of their host organisms over the majority of their history, but their potential for interspecific transfer could result in incongruencies between the two types of phylogenies (Stacey et al. 1986). However, very little is currently known about the frequency of interspecific transfer.

Finnegan (1985) has suggested a possible scenario for the origin and evolutionary history of transposable element families (Fig. 1). He suggests that a movable element could arise from host DNA sequences and this could, fortuitously, make use of endogenous cellular enzymes to replicate and insert copies

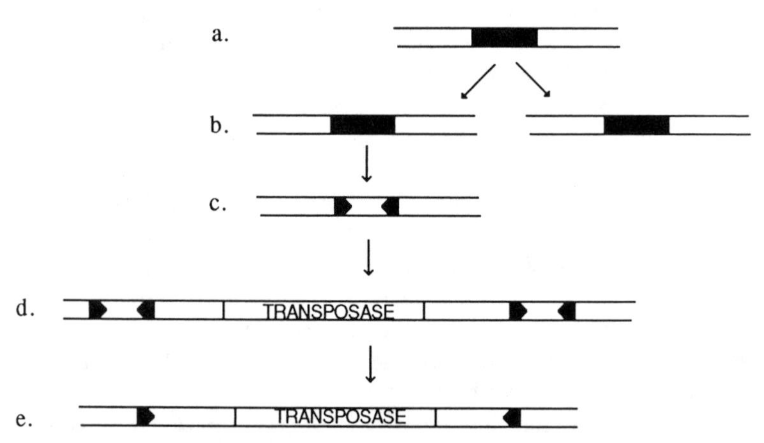

Fig. 1. A possible scenario for the origin and evolutionary history of a transposable element (after Finnegan 1985). A movable element **(a)** could arise from host DNA sequences and this could, fortuitously, make use of endogenous cellular enzymes to replicate and insert copies at new sites in the host genome **(b).** The accumulation of mutations which increase the efficiency and/or frequency of transposition of the primitive element might result in the creation of terminal repeats **(c).** A gene coding for an enzyme required for transposition might be incorporated into a more complex element by the flanking of such a gene with two primitive elements **(d).** Occasionally a transposition event might include both primitive elements and the intervening DNA coding for the transposase enzyme **(e).**

at new sites in the host genome. The human "Alu" family of repeats, which is unusual in lacking conserved repeated termini, might be an example of this primitive type of transposable element. It is expected that there would be strong natural selection for such a composite transposon. Examples of this evolutionary stage might be composite elements of *D. melanogaster* which are flanked by FB elements (Finnegan and Fawcett 1986). Mature elements would be subject to senescence if they were inactivated by mutations such as internal deletions. There are a number of examples of these types of degenerate stages such as the *P* element of *Drosophila* (O'Hare and Rubin 1983) in which deleted nonautonomous elements often constitute a majority of the elements of this family in active strains and are present as increasingly more degenerate types in many inactive strains. In the absence of any autonomous elements in the genome, immobilized elements are expected to be eventually lost since they can no longer replicate by means of transposition, but the removal of all members of a transposable element family is expected to take a long period of time (Charlesworth 1985).

TRANSPOSABLE ELEMENTS IN DROSOPHILA

Both the number and variety of known transposable element families in *D. melanogaster* are the largest of any species, probably because of the intensity with which it has been studied. These elements are an important constituent of the middle repetitive DNA and comprise at least 10% of the genome (Young 1979). Functional elements vary in size from about 2.6 to 8.8 kb. Depending on the family, they produce target site duplications of between 4 and 9 bp on insertion. Copy number per genome also tends to be characteristic for each family and varies between 6 and 80 with a mean of about 30. A detailed review of *D. melanogaster* transposable elements is provided by Finnegan and Fawcett (1986).

Despite the fact that *Drosophila* transposable elements are among the best studied of any organism, it is very difficult to decide on a classification that is meaningful, especially from an evolutionary standpoint. The most common way has been to classify these elements on the basis of of their structural properties. This provides some insights, but misses others. Another way is to focus on functional properties that different elements have in common, but the structural basis of these is often unclear. In the absence of an obvious choice, we have used a combination of both methods and have aimed to provide interesting examples rather than a systematic and exhaustive description.

Elements with conserved sequences. *D. melanogaster* contains a number of transposable element families whose sequences are highly conserved. These include the *copia*-like elements which characteristicly have long terminal repeats (LTRs). The length of these repeats is constant for each family, but varies between 250 and 500 base pairs (bp) in different families. With the exception of the *17.6* and *297* elements, there is no DNA sequence homology between families, but within families the elements are very similar to one another. The rate of transposition of these elements tends to be quite low (e.g., Montgomery and Langley 1983; Leigh-Brown and Moss 1987) and precise excision tends to be a very rare event. The high degree of sequence conservation appears to be related to the mode of transposition which involves a reverse transcriptase and an RNA intermediate. Unequal exchange has been postulated as a mode of copy number containment in this type of element (Langley et al. 1988).

From an evolutionary point of view, the *copia*-like elements are of considerable interest because of their structural similarity to the DNA proviruses of vertebrate retroviruses (Finnegan and Fawcett 1986). *Copia*-like elements and proviruses have a similar length and are both bounded by direct repeats of a few hundred base pairs. They are both flanked by target site duplications of a few base pairs at the site of insertion and are transcribed into full-length RNAs, starting in one LTR, and ending with the other. Both start with the sequence T-G and end with that of C-A. *Copia*-like elements, together with the *Ty* elements of yeast (Roeder and Fink 1983) have at least some, and sometimes all, of the functional sequences characteristic of proviruses. Comparison of amino acid sequences associated with reverse transcriptase and integrase activities coded by retroviruses and copia-like elements show a number of striking similarities (Fig. 2).

These and other similarities between proviruses and transposable elements provide a very strong argument that they are evolutionarily fairly closely related. Temin (1980) has argued that retroviruses have evolved from transposable elements, but Finnegan (1985) suggests that transposable elements might equally well be degenerate proviruses and that the two possibilities are not necessarily mutually exclusive.

Retroviral-like elements are the most numerous and widely distributed class of eukaryotic transposable elements. In *Drosophila* most morphologically detectable spontaneous mutations appear to result from the insertion of these

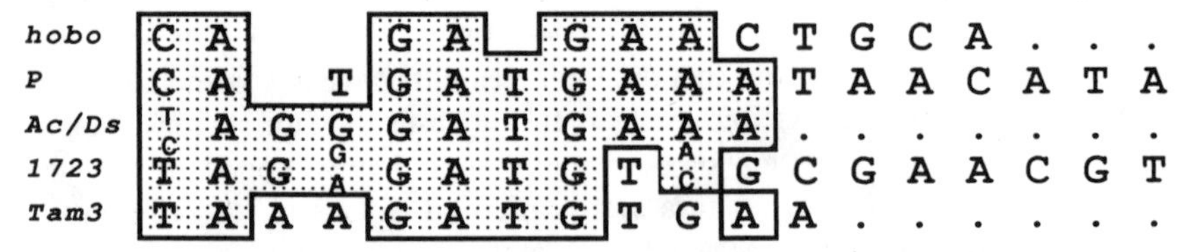

Fig. 2. Comparisons of amino acid sequences coded by vertebrate retroviruses and *Drosophila copia*-like transposable elements (after Finnegan and Fawcett 1986). Sequences are as follows: HTLV, human adult T-cell leukaemia virus; RSV, Rous sarcoma virus; MoMLV, Moloney murine leukaemia virus; 17.6 and *copia* are *Drosophila copia*-like transposable elements. Positions at which all five elements have identical or similar residues are boxed. Asterisks indicate positions with invariant residues.

elements (Sankaranrayanan 1988) and they appear to account for a significant portion of the allelic variation present in natural populations (Leigh-Brown 1983; Aquadro et al. 1986). Insertion mutants have been associated with new developmental and/or tissue-specific patterns of expression (Parkhurst and Corces 1985; Strand and McDonald 1989). Further, McDonald (1989) argues that this class of elements may play a role in the production of evolutionary novelties and in catalyzing the formation of new species, which may be especially pronounced under the special conditions existing in peripheral populations.

Unstable elements and those with long inverted repeats. There are two classes of elements in *D. melanogaster* that have long terminal repeats. These are the fold-back and TE elements. Fold-back elements have long, inverted repeat sequences of varying length and constitute about three percent of the total genome of *D. melanogaster.* TE elements are the largest known eukaryotic transposable elements and can be so large as to be detected as visible insertions in polytene chromosomes (Ising and Block 1981). TEs are highly unstable and may change their genetic content in association with transposition. The results of several sets of experiments suggest that any segment of DNA flanked by FB elements can transpose. This would explain the instability of TEs and their potential for producing large composite elements.

Elements that produce hybrid dysgenesis. The term "hybrid dysgenesis" was first introduced (Kidwell et al. 1977) to describe a syndrome of unusual phenotypic traits that were induced by outcrossing males from a number of *D. melanogaster* strains with females from long-established laboratory strains. Two families of transposable elements, the *P* elements (Bingham et al. 1982) and the *I* elements (Bucheton et al. 1984) are responsible for the P-M and the I-R systems of hybrid dysgenesis, respectively. Subsequently, *hobo,* a third family of transposable elements has been described whose activation produces a number of phenotypic manifestations that have some of the characteristics shared by the P-M and I-R systems of hybrid dysgenesis (reviewed by Blackman and Gelbart 1989). The *P* and *hobo* elements share certain structural features with one another and with the *1723* element of *Xenopus laevis*, the *Ac/Ds* elements of maize, and the Tam3 element of *Anntirhinum majus* (Streck et al. 1986). All of these elements have short inverted repeats which share some sequence similarity in their termini (Fig. 3).

Hybrid dysgenesis was originally described on the basis of the unusual phenotypic properties and modes of inheritance which were observed to be associated with the P-M and I-R systems. These two systems share several common features. The most important of these are the associated induction of a number of dysgenic traits with nonreciprocal, hybrid manifestation, tissue specificity and environmental dependence. Reciprocal cross differences are due to the maternal inheritance of an element-encoded repressor. In the P-M system, this represents one aspect of an autoregulation system by which some

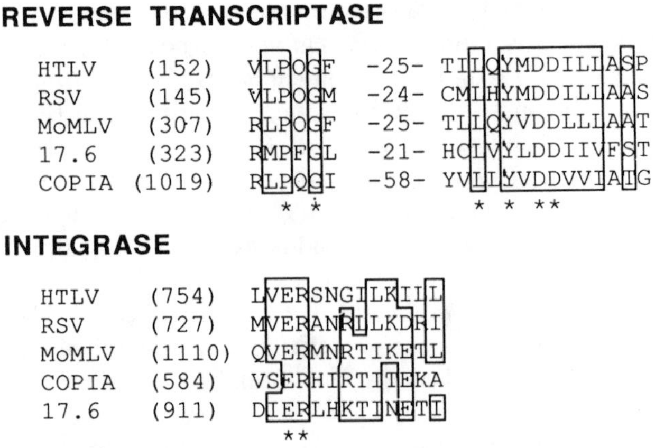

REVERSE TRANSCRIPTASE

```
HTLV    (152)   VLPQGF  -25-  TILQYMDDILLASP
RSV     (145)   VLPQGM  -24-  CMLHYMDDILLAAS
MoMLV   (307)   RLPQGF  -25-  TLLQYVDDLLLAAT
17.6    (323)   RMPFGL  -21-  HCLVYLDDIIVFST
COPIA  (1019)   RLPQGI  -58-  YVLLYVDDVVIATG
                  *  *          *  *  **
```

INTEGRASE

```
HTLV    (754)   LVERSNGILKIIL
RSV     (727)   MVERANRILKDRI
MoMLV  (1110)   QVERMNRTIKETL
COPIA   (584)   VSERHIRTITEKA
17.6    (911)   DIERLHKTINETI
                  **
```

Fig. 3. Comparisons of terminal repeats of transposable elements that show some structural similarities from diverse organisms (after Streck et al. 1986). *hobo* and *P* are *D. melanogaster* elements. The *Ac/Ds* element system is found in maize. *Tam3* and *1723* are elements found in *Antirrhinum majus* and *Xenopus laevis*, respectively. The left end of each sequence shown is the element's terminus. All of the elements produce 8-base duplications of the target sequence. Both nucleotides are shown for positions in which the two ends can vary.

classes of internally-deleted, nonautonomous, *P* elements regulate their family copy number (see Engels 1989 for a recent review of *P* elements).

The simultaneous occurrence of a number of associated genetic traits is characteristic of active transposable element systems in a wide variety of species (reviewed in Shapiro 1983). The association of multiple traits appears to reflect their coincident induction and the multiple ways that host sequences can be changed due to transposition, excision and other activities of mobile element systems. However, in hybrid dysgenesis, both the repertoire, and possibly the frequency, of dysgenic traits appears to be enhanced in comparison with those of mobile systems in general. In addition to high frequencies of insertion and excision mutations, transmission ratio distortion, male recombination and chromosomal aberrations, hybrid dysgenesis determinants can produce variable levels of temperature-sensitive sterility.

An example of the dramatic enhancement of mutation frequency that may be associated with hybrid dysgenesis is provided by the X-linked singed-weak allele. This allele results from the insertion of two deleted elements in a head-to-head orientation (Roiha et al. 1988). It mutates to near wild type, or to an extreme singed allele, at inordinately high rates, sometimes in excess of 50% (Engels 1979), by excision of one or the other of the two *P* elements.

Hybrid dysgenesis in the P-M and I-R systems is normally restricted to the germ line. Although somatic and germ line cells share the same genetic complement of *P* and *I* elements, in somatic cells these elements are usually quite stable (Bregliano and Kidwell 1983). The germline specificity of *P* element transposition is due to differential splicing of the complete *P* element transposase transcript in germ line and soma (Laski et al. 1986). Regulation of gene expression by this splicing mechanism is estimated to occur in approximately 5% of chromosomal genes (Bingham et al. 1988). From an evolutionary standpoint, the restriction of transposition to the germline reduces the deleterious effects of somatic transposition on the host. This is expected to favor the maintenance of the elements in the host genome.

THE UNLIKELIHOOD OF HYBRID DYSGENESIS AS A REPRODUCTIVE ISOLATING MECHANISM

The similarities between the types of sterility induced by hybrid dysgenesis and those observed in the progeny of interspecific and intersubspecific crosses early suggested that hybrid dysgenesis might be an important mechanism in the evolution of postmating isolation. Several relatively simple models have been proposed for the evolution of speciation involving transposable elements (e.g., Ginzburg et al. 1984; Rose and Doolittle 1983), but they did not include the complexity of element regulation and were, therefore, somewhat unrealistic. However, Kreiber and Rose (1986) conclude that there is little, if any, evidence for the production of speciation by any simple molecular mechanism, including hybrid dysgenesis. A closer examination of the dynamics of hybrid dysgenesis determinants suggests that reproductive isolation is unlikely to result from the sterility which they induce. Because of the highly invasive nature of known hybrid dysgenesis determinants and the conditional and partial nature of the induced hybrid sterility and the evolution of self regulation, it appears to be unlikely on theoretical grounds that hybrid dysgenesis can act directly as a post-mating reproductive isolating mechanism. Another approach to answering the question of possible mobile element involvement in the evolution of postmating reproductive isolation has been to ask whether there are any indications of hybrid dysgenesis when closely-related species are crossed. Many crosses between closely related species and subspecies of *Drosophila* show reciprocal differences in sterility induction and other superficial similarities to hybrid dysgenesis. However, experiments to test the involvement of transposable elements have so far produced negative results (e.g., Hey 1985) and other mechanisms, such as X-chromosome-autosome incompatibilities (e.g., Coyne 1984) and sexual selection leading to premating isolation (Kaneshiro and Boake 1987), have been demonstrated to account for reciprocal differences in some instances. Although there appears to be little support for a direct role of hybrid dysgenesis in speciation, an indirect role in reproductive isolation cannot, however, be ruled out at the present time. For example, hybrid dysgenesis-induced sterility might serve to reinforce other isolating mechanisms under some circumstances. Chromosomal aberrations and new behavioral mutants induced by hybrid dysgenesis might lead, respectively, to post-mating and pre-mating reproductive isolation.

SPECIES DIFFERENCES IN THE MIDDLE REPETITIVE FRACTION OF THE GENOME

Several studies have shown that within the genus *Drosophila* there is considerable variability in the amount and distribution of dispersed repetitive DNA, even among closely related species. For example, one study suggested that the chromosomes of *D. melanogaster* carry approximately three times as much repetitive DNA as those of the sibling species *D. simulans.* There is at least a two-fold difference in their dispersed and repetitive DNA contents (Dowsett and Young 1982, Hey and Eanes 1990). It has been speculated that these differences may influence the relative mutation rates of these species and/or contribute to their reproductive isolation.

Along the same lines Dowsett (1983) concluded that closely related species can contain different libraries of repetitive elements in their genomes. He examined six closely related species of the *D. melanogaster* species subgroup. Of the 28 repetitive families isolated from the *D. melanogaster* genome, only two were present in the other five species. Similarly, another species, *D. erecta,* possesses another set of repetitive families which are absent from the *D. melanogaster* genome.

Examination of the middle repetitive DNA of *D. algonquin,* a species belonging to the obscura subgroup of the *melanogaster* species group, showed that this species differs markedly in its transposable element properties from that of *D. melanogaster* (Hey 1989). Four element families in *D. algonquin* varied in copy number from 59 to 333 with an occupancy rate per site that varied from 0.64 to 0.75. The values for the sibling species *D. affinnis* were very similar. However, they are both in sharp contrast to *D. melanogaster,* which has at least 30 different families with copy numbers varying from about 5 to 100 and a much lower occupancy rate per site.

Thus, individual middle repetitive families appear to be highly unstable in *Drosophila* genomes over short periods of evolutionary time and closely related species may have widely varying amounts and patterns of distribution of these families. Considerable care must therefore be taken in the interpretation of results and in avoiding unwarranted generalizations.

DISTRIBUTION OF MOBILE ELEMENTS IN THE GENUS DROSOPHILA

Transposable elements can be transmitted by either vertical (mating-dependent) or horizontal (mating-independent modes). The two modes have different expectations with respect to distribution patterns (Stacey et al. 1986). Vertical transmission leads to continuous distributions with distribution patterns of the element that are congruent with host phylogeny. Horizontal transmission tends to produce discontinuous or patchy distributions with sequence conservation of the element not necessarily being related to the phylogeny of the host.

A start has been made to looking at the distribution of different mobile element families within the genus *Drosophila.* Probes specific for a number of different families from a few species have been hybridized to genomic DNA of a number of other species. Studies of the distribution patterns of several *D. melanogaster* elements provide quite variable results. The *hobo* element has the most limited distribution. It is restricted to the melanogaster and montium subgroups of the *melanogaster* species group (S. B. Daniels and I.A. Boussy, unpublished results) have very limited distributions. The I element is somewhat less restricted, being present in all except one species of the melanogaster subgroup (Stacey et al. 1986). *Copia, gypsy* and *F* elements are widespread in the genus, but there are large differences between species with respect to their similarity to the *D. melanogaster probes* (Martin et al. 1983; Stacey et al. 1986). The distribution of copia and the conservation pattern of the gypsy element do not support a simple model of a single ancient origin of these elements, followed by mating-dependent transmission, but are consistent with the occurrence of a few horizontal transmission events between reproductively isolated species.

The distribution of *P* sequences within the genus *Drosophila* is quite patchy and, in many cases, not congruent with phylogenetic groupings (Lansman et al. 1985; Daniels and Strausbaugh 1986; Stacey et al. 1986). *P*-homologous sequences are widely distributed among all the species groups that comprise the subgenus Sophophora although, interestingly, they are not found in the species most closely related to *D. melanogaster,* including its sibling species *D. simulans* (Brookfield et al. 1984). However, most species of the distantly-related willistoni subgroup show homology to the *D. melanogaster P* element, as do two species in the nondrosophilid families Trixoscelididae and Opomyzidae (Anxolabehérè et al. 1987, S.B. Daniels et al., in preparation). *P* elements appear to have had a long-standing presence in many *Drosophila* species, but they appear to be no longer active and exist as deleted, presumably relic, sequences (Daniels and Strausbaugh 1986).

Overall, the available evidence on *P* element distribution is consistent with inter-specific transfer events having taken place at varying times in the history of the genus. This pattern is consistent with the hypothesis that *P* elements have only recently been introduced into the *D. melanogaster* genome, presumably by horizontal transmission from some as yet unknown source (Anxolabehérè et al. 1988; Kidwell 1983). The recent determination that a complete *P* element from *D. willistoni* has a nucleotide sequence essentially identical to that of complete *D. melanogaster P* elements (S. B. Daniels and K. R. Peterson, unpublished results) provides additional support for the hypothesis of interspecific transfer and strongly suggests that *D. willistoni* was the donor species.

EVOLUTIONARY SIGNIFICANCE

A question of considerable interest is to what extent transposable elements are or have been maintained by virtue of benefits which they confer on their host organisms, and to what extent they are maintained by their ability to over-replicate relative to their host sequences. The possibility that transposable elements represent a type of selfish DNA was cogently argued by Doolittle and Sapienza (1980) and Orgel and Crick (1980). Subsequently the term *parasitic DNA* has tended to replace the term *selfish DNA* because of the anthropomorphic connotations of the latter term. More recently it has also been argued (e.g., Ajioka and Hartl 1989) that currently available evidence is most consistent with a model of parasitic DNA, at least in the short term. This evidence includes the distribution patterns of transposable elements in present day natural populations and the absence of fixed insertion sequences in host genes which might suggest that there has been selection for favorable mutations (Montgomery and Langley 1983). However, in some prokarytotes, certain temperate bacteriophage and insertion sequences in chemostats have been shown to have a favorable effect on host fitness. Also, some mathematical models (Condit et al. 1988) are not consistent with a largely parasitic role of transposable elements in prokaryotes.

When a long-term evolutionary perspective is considered, Campbell (1981) has argued that the parasitic DNA model is insufficient to explain the presence of these sequences over long periods of evolutionary time. This is because the cost to the organism of maintaining the sequences is likely to lead to their eventual elimination if their role is solely parasitic. A major way that transposable elements could confer advantages to their hosts is by either mobilizing or changing the expression of host genes or by acting as potent mutagenic agents. Most mutations would be expected to be deleterious but, as with other mutagens, a small subset of them might be advantageous to the host. Transposable element-induced beneficial mutations or rearrangements would be spread by selection, thus contributing to the maintenance of the transposable elements themselves in the population.

Several data sets (e.g., Montgomery and Langley 1983; Leigh-Brown and Moss 1987) indicate that *D. melanogaster* transposable elements have a very low occupancy per site. This is inconsistent with a model of maintenance by direct selection of favorable mutations. However, the situation in *D. melanogaster* may not be typical of that in other species. For example, in *D. algonquin* (Hey 1989), the number of transposable element families is considerably reduced relative to that in *D. melanogaster* and the occupancy rate per site is very much higher. It is therefore premature to make general conclusions on the basis of the currently rather limited data.

An alternative possibility is that transposable elements are maintained over long periods of evolutionary time because they confer advantages at the level of the population rather than that of the individual. The potential for high mutability, possibly in response to environmental change, is a good example of a transposable element property that might be selected in this way. Species selection models must deal with the problems that are associated with group selection models in general (Maynard-Smith 1976), particularly the problem of how elements can spread initially in the face of deleterious effects that they produce. However, *D. melanogaster* provides a well-documented example of element invasion of a species in the face of deleterious effects which might be far more severe than those expected from the induction of unfavorable mutations. Extensive data on the historical and geographical distribution of *P* elements in *D. melanogaster* (Kidwell 1983; Anxolabéhère et al. 1987) indicate that *P* elements have invaded this species worldwide during the last 40 years, despite an expected severe reduction in fitness of hybrids. The ability of *P* elements to over-replicate and the ease with which contemporary *D. melanogaster* can migrate long-distance are among the factors which may have allowed this recent worldwide invasion. *P* elements have also been shown to invade experimental populations despite their initial low frequency and a high frequency of dysgenesis-associated sterility under restrictive environmental conditions (Kiyasu and Kidwell 1984). The behavior of *P* elements suggests that the difficulties associated with group selection models may be more easily overcome by some transposable elements than by more conventional genes.

Ajioka and Hartl (1989) have pointed out that the relative importance of different forces which bear on the evolution of transposable elements today may vary considerably from those that acted at the time of their origin and early evolution. For example, the appearance of new mutations may have been a limiting factor during some earlier stages of host evolution, but be less important in more highly evolved contemporary organisms.

If there is an overall generalization, it is that the evolution of transposable elements is intimately related to that of their hosts, whether or not that relationship is essentially parasitic or mutually beneficial to each. There are probably many different ways that hosts and parasites or symbionts adapt to one another, depending on the individual characteristics of each. Probably one of the most satisfactory ways to view the evolution of transposable elements is in terms of sequential alternate phases of parasitic and mutually symbiotic relationships (Syvanen 1984). The relative length of different phases might vary considerably, with parasitic relationships predominating, particularly in more recently evolved eukaryotes.

CONCLUSIONS

A number of insights have been gained from the discovery and characterization of transposable elements. These have lead to changes in our concepts of genes and genomes which are among the most interesting that have emerged from the recent application of recombinant DNA techniques to the study of biology. Transposable elements have a widespread distribution in living organisms and seem likely to have had an ancient origin in the history of life.

Families of mobile sequences from a wide diversity of organisms share a number of common structural and functional properties, but are also characterized by a number of differences. They can mediate almost every known type of mutational and recombinational change in the DNA of their host organisms. Possible scenarios for the origin and course of evolution of transposable elements are described. A number of examples from *Drosophila* are provided to illustrate the diversity of structures, variety of phenotypic effects and distributions found among mobile sequences. Hypotheses for the maintenance of these elements in host genomes are also discussed. The most likely way that *Drosophila* transposable elements could have evolutionary significance for their host organisms seems to be through their mutagenic properties and ability to act as portable sites of homology for recombination. The evolution of transposable elements is intimately related to that of their hosts. Their evolutionary significance is discussed in terms of sequential, alternate phases of parasitic and mutually symbiotic relationships, with the relative length of different phases varying over evolutionary time. Current evidence is consistent with predominantly parasitic relationships, particularly during relatively recent stages of the evolution of *Drosophila*.

REFERENCES

Ajioka JW, Hartl D (1989): Population dynamics of transposable elements. In Berg DE, Howe M (eds): "Mobile DNA," Washington DC: Amer Soc Microbiol Publ.

Anxolabéhère D, Kidwell MG, Périquet G (1988): Molecular characteristics of diverse populations are consistent with the hypothesis of a recent invasion of *Drosophila melanogaster* by mobile P elements. Mol Biol Evol 5:252–269.

Anxolabéhère D, Nouaud D, Périquet G (1987): Evolutionary genetics of the *P* transposable elements in *Drosophila melanogaster* and in the Drosophilidae family. Life Science Advances C1:37–46.

Aquadro CF, Deese SF, Bland MM, Langley CH, Laurie-Ahlberg CC (1986): Molecular population genetics of the alcohol dehydrogenase gene region of *D. melanogaster.* Genetics 114:1165–1190.

Bingham PM, Kidwell MG, Rubin GM (1982): The molecular basis of P-M hybrid dysgenesis: The role of the *P* element, a *P* strain-specific transposon family. Cell 29:995–1004.

Bingham PM, Chou TB, Mims I, Zachar Z (1988): On/off regulation of gene expression at the level of splicing. Trends Genet 4:134–138.

Blackman R, Gelbart W (1989): The transposable element *hobo* of *Drosophila melanogaster.* In Berg DE, Howe M (eds): "Mobile DNA." Amer Soc Microbiol Publ.

Bregliano JC, Kidwell MG (1983): Hybrid dysgenesis determinants. In Shapiro JA (ed): "Mobile Genetic Elements." New York: Academic Press, p 363.

Brookfield JFY, Montgomery E, Langley CH (1984): Apparent absence of transposable elements related to the *P* elements of *D. melanogaster* in other species of *Drosophila.* Nature 310:330–332.

Bucheton A, Paro R, Sang HM, Pelisson A, Finnegan DJ (1984): The molecular basis of I-R hybrid dysgenesis: Identification, cloning and properties of the I factor. Cell 38:153–163.

Campbell A (1981). Evolutionary significance of accessory DNA elements in bacteria. Annu Rev Microbiol 35:55–83.

Chao L, Vargas C, Spear BB, Cox EC (1983): Transposable elements as mutator genes in evolution. Nature 303:633–635.

Charlesworth B (1985): The population genetics of transposable elements. In Ohta T, Aoki KI (eds): "Population Genetics and Molecular Evolution." New York: Springer-Verlag, pp 213–232.

Condit R, Stewart FM, Levin BR (1988): The population biology of bacterial transposons: A priori conditions for maintenance as parasitic DNA. Am Naturalist 132:129–147.

Coyne JA (1984): Genetic basis of male sterility in hybrids between two closely related species of *Drosophila.* Proc Natl Acad Sci USA 81:4444–4447.

Daniels SB, Strausbaugh LD (1986): The distribution of P-element sequences in *Drosophila:* The *willistoni* and *saltans* species groups. J Mol Evol 23:138–148.

Di Nocera PP, Casari G (1987): Related polypeptides are encoded by *Drosophila* F elements, I factors and mammalian L1 sequences. Proc Natl Acad Sci USA 87:5843–5847.

Doolittle WF, Sapienza C (1980): Selfish genes, the phenotype paradigm and genome evolution. Nature 284:601–603.

Dowsett AP (1983): Closely related species of *Drosophila* can contain different libraries of middle repetitive DNA sequences. Chromosoma 88:104–108.

Dowsett AP, Young MW (1982): Differing levels of dispersed repetitive DNA among closely related species of *Drosophila.* Proc Natl Acad Sci USA 79:4570–4574.

Engels WR (1979): Extrachromosomal control of mutability in *Drosophila melanogaster.* Proc Nat Acad Sci USA 76:4011–4015.

Engels WR (1989): *P* elements in *Drosophila.* In Berg DE, Howe M (eds): "Mobile DNA." Washington, DC: Amer Soc Microbiol Publ, pp 437–484.

Finnegan DJ (1985): Transposable elements in eukaryotes. Int Rev Cytol 93:281–326.

Finnegan DJ, Fawcett DH (1986): Transposable elements in *Drosophila melanogaster.* In Maclean N (ed): "Oxford Surveys on Eukaryotic Genes." Vol 3, Oxford: Oxford Univ Press, pp 1–62.

Fitzpatrick B, Sved JA (1986): High levels of fitness modifiers induced by hybrid dysgenesis in *D. melanogaster.* Genet Res 48:89–94.

Ginzburg L, Bingham P, Yoo S (1984): On the theory of speciation induced by transposable elements. Genetics 107:331–341.

Hey J (1985): Speciation via hybrid dysgenesis is unlikely. Genetics 110:85.

Hey J (1989): The transposable portion of the genome of *Drosophila algonquin* is very different from that of *D. melanogaster.* Mol Biol Evol 6:66–79.

Hey J, Eanes WF (1990): An examination of transposable element identity and copy number in the euchromatin of *Drosophila melanogaster.* Genetics (in press).

Hickey DA (1982): Selfish DNA: A sexually-transmitted nuclear parasite. Genetics 101: 519–553.

Ising G, Block K (1984): A transposon as a cytogenetic marker in *Drosophila melanogaster*. Mol Gen Genet 196:6–16.

Kaneshiro KY, Boake CRB (1987): Sexual selection and speciation: Issues raised by Hawaiian drosophilids. Trends Ecol and Evol 2:207–211.

Kaplan N, Darden T, Langley C (1985): Evolution and extinction of transposable elements in Mendelian populations. Genetics 109:459–480.

Kidwell MG (1983): Evolution of hybrid dysgenesis determinants in *Drosophila melanogaster.* Proc Nat Acad Sci USA 80:1655–1659.

Kidwell MG, Kidwell JF, Sved JA (1977): Hybrid dysgenesis in *Drosophila melanogaster:* A syndrome of aberrant traits including mutation sterility and male recombination. Genetics 36:813–833.

Kiyasu PK, Kidwell MG (1984): Hybrid dysgenesis in *Drosophila melanogaster:* The evolution of mixed P and M populations maintained at high temperature. Genet Res, Camb 44:251–259.

Kreiber M, Rose MR (1986): Molecular aspects of the species barrier. Ann Rev Ecol Syst 17:465–485.

Langley CH, Montgomery E, Hudson R, Kaplan N, Charlesworth B (1988): On the role of unequal exchange in the containment of transposable element copy number. Genet Res 52:223–235.

Lansman RA, Stacey SN, Grigliatti TA, Brock HW (1985): Sequences homologous to the *P* mobile element of *Drosophila melanogaster* are widely distributed in the subgenus Sophophora. Nature 318:561–563.

Laski FA, Rio DC, Rubin GM (1986): Tissue specificity of *Drosophila P* element transposition is regulated at the level of mRNA splicing. Cell 44:7–19.

Leigh EG Jr (1973): The evolution of mutation rate. Genetics Suppl 73: 1–18.

Leigh-Brown AJL, Moss JE (1987): Transposition of the *I* element and *copia* in a natural population of *Drosophila melanogaster*. Genet Res 49:121–128.

Martin G, Wiernasz D, Schedl P (1983): Evolution of *Drosophila* repetitive-dispersed DNA. J Mol Evol 19:203–213.

Maynard Smith J (1976): Group selection. Quart Rev Biol 51:277–283.

McDonald JF (1989): The potential evolutionary significance of retroviral-like transposable elements in peripheral populations. In Fontdevilla A (ed): "Genetics of Unstable Populations." New York: Springer-Verlag.

Montgomery EA, Langley CH (1983): Transposable elements in Mendelian populations. II. Distribution of three *copia*-like elements in a natural population of *Drosophila melanogaster.* Genetics 104:473–483.

O'Hare K, Rubin GM (1983): Structures of *P* transposable elements and their sites of insertion and excision in the *Drosophila melanogaster* genome. Cell 34:25–35.

Orgel LE, Crick FHC (1980): Selfish DNA: The ultimate parasite. Nature 284:604–606.

Parkhurst SM, Corces VG (1985): *Forked, gypsys*, and suppressors in *Drosophila*. Cell 41:429–437.

Roeder GS, Fink GR (1983): Transposable elements in yeast. In Shapiro A (ed): "Mobile Genetic Elements." New York: Academic Press, pp 299–328.

Roiha H, Rubin GM, O'Hare K (1988): *P* element insertions and rearrangements at the singed locus of *Drosophila melanogaster.* Genetics 119:75–83.

Rose MR, Doolittle WF (1983): Molecular biological mechanisms of speciation. Science 220:157–162.

Sankaranarayanan K (1988): Mobile genetic elements, spontaneous mutations and the assessment of genetic radiation hazards in man. In Lambert ME, McDonald JF, Weinstein IB (eds): "Eukaryotic Transposable Elements as Mutagenic Agents." Cold Spring Harbor (NY): Cold Spring Harbor Laboratories, pp 319–336.

Shapiro J (1983): "Mobile Genetic Elements." New York: Academic Press.

Stacey SN, Lansman RA, Brock HW, Grigliatti TA (1986): Distribution and conservation of mobile elements in the genus *Drosophila.* Mol Biol Evol 3:522–534.

Strand DJ, McDonald JF (1989): Insertion of a *copia* element 5′ to the Drosophila melanogaster alcohol dehydrogenase gene (adh) is associated with altered developmental and tissue–specific patterns of expression. Genetics 121:787–794.

Streck RD, MacGaffey JE, Beckendorf SK (1986): The structure of *hobo* transposable elements and their insertion sites. EMBO J 5:3615–3623.

Syvanen M (1984). The evolutionary implications of mobile genetic elements. Annu Rev Genet 18:271–293.

Temin HM (1980): Origin of retroviruses from cellular moveable genetic elements. Cell 21:599–600.

Young MW (1979): Middle repetitive DNA: A fluid component of the *Drosophila* genome. Proc Natl Acad Sci USA 76:6274.

New Perspectives on Evolution, pages 155-163

Gene Structure and Evolutionary Theory

WALTER GILBERT
Department of Cellular and Developmental Biology, Harvard University, Cambridge, Massachusetts 02138

Modern genes are made up of introns and exons. To what extent can we understand that gene structure and follow it back to the beginning of evolution?

In general, the genes of higher organisms are broken up along the DNA into the sequences that are expressed in the final product, which we call exons, and sequences that separate those regions along the DNA, which we call introns. The most general property of those two sequences is that the introns are very much larger than the exons, extending roughly some ten to twenty times farther along the DNA. A second property is that the distribution of exon sizes is peaked at around 40 or 50 amino acids—150 bases—whereas the distribution of intron sizes is very broad and the longest introns extend out to hundreds of kilobases.

If we compare the sequences for homologous genes in different organisms, the exon sequences are for the most part conserved, since they are related to the protein sequences, whereas the introns drift rapidly in sequence and in length: drifting as rapidly as the third positions of the codons, drifting as though they are DNA's that are unconstrained in function. What can the introns be doing? The hypothesis that I'd like to suggest to you is that the role of the intron sequences is simply to shuffle the exons into novel combinations. This is not to suggest that the intron sequences are preserved through evolution because they are going to be used in the future for that purpose. It is simply the argument that if there is no great pressure to eliminate extra DNA—the organism is not under DNA-content pressure—then just the fact that the intron sequences were used in the last episode of gene creation means that we will see them still present today. The argument that the intron sequences increase the recombination rate between the exons is simply the following. Think of two exons that one wishes to combine into a novel gene structure. A recombination in the intron DNA that can combine the two exons can occur anywhere within 10,000 bases after one of the exons, anywhere within 10,000

An edited transcript of Walter Gilbert's talk in the session on "Impact of Recent Advances in Molecular Biology I" during the second day of the symposium on New Perspectives on Evolution, April 19, 1989.

bases before the other, to produce a structure containing both separated by a 10,000 bp intron. Such recombination is a factor of a hundred million faster than it would be if that recombination were constrained, so as to attach just the end of one coding sequence to the end of the other coding sequence.

In general, if we look at genes in the vertebrates that correspond to proteins that have a clearly repeated or multimeric structure, we find that those structures contain repeated exons separated by introns. The gene for collagen from the chicken, analyzed by de Crombrugghe and his coworkers, scattered along some 50 kilobases of DNA, consists of a 50-fold repeat of a small exon coding for 18 amino acids—exactly the sort of structure we would expect, had that gene been built up by recombination from the small unit, using the intron/exon structure.

The immunoglobulin superfamily contains many examples of a single exon that bears a particular structural motif, the immunoglobulin fold, repeated over and over again, in single copies, in two copies, in four copies, in five copies, and in arbitrary distributions. We would expect, if introns had been used to recombine exons, that we would find examples of exon shuffling, with the same exons serving as units in dissimilar genes. The most dramatic example was found a few years ago, in the comparison between the structure of the LDL receptor, worked out by Brown and Goldstein and their collaborators, and the epidermal growth factor precursor, worked out by the group at Chiron. There is a 40-amino-acid-long exon that appears three times in the LDL receptor, that appears again three times in the EGF precursor, and that appears again in the blood clotting factors. These unrelated proteins contain a structural motif, born on an exon, that appears in each gene.

This is the basis for the argument that if we look at the late genes—genes that have arisen late in evolution, during the vertebrate radiation—they have an intron/exon structure where one can see that complicated genes have been put together by the introns and that exons have been shuffled. Can we extend this conception of the structure of genes back further in time? Several years ago my laboratory looked at a specific enzyme, triose phosphate isomerase, a very ancient protein, whose amino acid sequence is conserved across procaryotes and eucaryotes to such a degree that we infer that its three-dimensional structure is the same everywhere. This enzyme reached what Jeremy Knowles' group calls "enzymatic perfection" far back in the mists of time before the division of the procaryotes and the eucaryotes. Such a gene has a single continuous structure along the DNA in bacteria; what do we find in complex organisms? If the introns had been added at some time in evolutionary history, then we might expect that gene structure not broken be up at all in modern organisms. If the introns had existed from the very beginning of evolution, then we might expect that gene to still retain the signs of that original structure. We now know that the genes for the glycolytic enzymes are, in general, broken up extensively into an intron/exon structure. For triose phosphate isomerase, we discovered six introns in the protein in chickens. We then looked for another complex eucaryote that might also have a pronounced intron/exon structure but which would be very distant. We examined the same gene in corn, and Mark Marchioni found eight introns in that gene. The striking aspect of that comparison is that five of those introns are at the same sites in the two genes; one of the introns is three amino acids over; and two of the introns in plants occur at positions at which no intron appears in the animal. What do we make of this? The coincidence in position of these introns is not what we would expect had the introns been added after the division of the plant lines from the animal lines. Rather, the introns must have existed in the primeval organism that led down to both the plants and the animals. The interpretation is that five introns are at the original positions, one intron has moved three amino acids over (the movement of an intron can be achieved with sequential single mutations, because a mutation that destroys a splicing site is often immediately compensated for by the revelation of some cryptic splicing site that can be used instead of the original one), and two introns were lost down one line and not the other. This is about as far back as one can go convincingly to argue that the intron structure existed in single-cell organisms before the division of plants and animals, a time on the order of a billion years ago. If the introns existed then, then when we look at the simpler eucaryotes—at genes in *Drosophila* or genes in *C. elegans*—where we find no introns, we can infer that those represent examples of loss of introns.

Now I'd like to generalize that thought to the idea that the first gene structures were made up of small exons. The extreme position is that the first exons were 15 to 20 amino acids long, small polypeptides with some functional form and role. Ten years ago, one thought that small polypeptides did not have structures in solution, but the physical chemistry of recent years supports a new attitude: that one can find examples of short polypeptides, 15 to 20 amino acids long, which have clearly defined structures in solution. Buzz Baldwin has studied an example of such a polypeptide that forms an alpha helix in solution and Lerner and his colleagues have also studied several.

The picture that emerges is that during the process of evolution genes are created out of combinations of small exons, the exons are shuffled to make new combinations, and furthermore, over evolutionary time, introns are lost leading to the gradual creation of more and more complex exons. At least one process occurs, retroposition, in which an RNA copy splices out the introns and then is copied back into DNA and inserted into the chromosome. That process creates pseudogenes, but if such structures are inserted into a transcription unit, that reinserted element now serves as an exon in a more complicated protein. There are now several examples of such gene structures. The beta–adrenergic receptors are related to the opsins, but the opsins have a full intron/exon structure while the adrenergic receptors have no introns.

This is a view of evolution in which the processes are basically recombination of introns, the sliding and drift of introns, and the loss of introns by a variety of processes: Exact deletion, which we expect to be extremely rare, inexact deletion, which is still quite rare, and the retroposon processes, which occur at quite a reasonable evolutionary rate (about once each hundred million years). A major prediction of this theory is that proteins will turn out be made up of separate modules, elements of folding, and that ultimately these modules will be correlated with the exons. An example of this process is in the immunoglobulin superfamily. Most have a typical exon, bearing about 112 amino acids. However, CD4, a surface molecule on T cells, has the immunoglobulin fold broken up into two pieces. N-CAM also has the immunoglobulin fold broken up into pieces. These are examples of older structures, the pieces of which are more closely related to the mini-exons that were assembled to make the complicated immunoglobulin fold.

We look upon the original cell—which came into existence some three to four billion years ago—as having an intron/exon structure, a complicated splicing pattern, which was refined over evolutionary time down a line that leads directly to the eucaryotic nucleus and finally to multicellular organisms. In this picture, the lines down to the procaryotes and the archebacter are lines that lost introns because the organisms were specialized for rapid growth in such a way as to put pressure on their DNA content. The invasion of offspring of these partially degenerate organisms into cells, to produce the mitochondria and chloroplasts, created what we think of as the modern eucaryotic cell about a billion and a half to two billion years ago, but the structure of the eucaryotic nucleus, with its introns and exons as seen in the most complicated eucaryotes, is just a remnant of the original nuclear structure in the original cells.

This picture offers a way of solving a numerical paradox critical to evolution that was alluded to by Hartl yesterday. How does evolution make a complicated protein? Do we search out the entire sequence space made up of all the possible sequences of amino acids? For a 200-amino-acid protein, do we ever search out the space of 20^{200} sequences? Of course the answer to that question has to be no, because there is no way in the visible universe, and in the length of time the universe has existed, to search so many sequences. There are several ways around this paradox. The exon idea is one such. It gets around this paradox by the simple expedient of using shapes drawn from sequences of 15 to 20 amino acids in length: only 20^{15} to 20^{20} possible amino acid sequences. That's a few tons of material. One can simplify the evolutionary problem further by saying that a set of reasonably useful shapes characterize those polypeptides that actually have some conformation; that's a smaller number of molecules. For example, this set might be on the order of a million molecules. How does one make complicated sequences out of those million molecules? One finds a first, by searching once through the million molecules to find a function. One then adds a second exon by searching through a million molecules again to find a somewhat better function, and then one adds to that again, and again. To make a 200-amino-acid-long protein only requires a search through 10^7 different possibilities. This is a tremendous reduction of the search problem, that follows automatically if the proteins are made of modular subunits, each with a certain function. The consequence is that the proteins that evolution would have come up with only represent a tiny fraction of the possible shapes that could be constructed from amino acid sequences. Therefore, the prediction is that there are completely different ways of solving every enzymatic problem by shapes that nature has never gone near.

How many shapes are actually used to make proteins? We have attempted such a calculation to estimate the size of the universe of exons from which all the proteins are derived. If the known gene structures represented some random selection of exons out of a universe of possible exons, then we could examine the gene structures and ask, how many examples of exon shuffling and how many repeats are there? If we have selected M exons out of a universe of size N, then the number of pairs of exons is M (M-1)/2. The probability that any of those pairs will match is just the reciprocal of the size of the underlying universe. So M (M-1)/2 N is an expression for the number of repeats we expect. To do that calculation, we went to the database to find a list of all the exons. That turns out not to be easy, even though the database is computerized. One has to work over the data base by hand to generate a database of exons. The first problem is that the selection of known genes is not random; it is biased by the ways in which people have looked at genes. For instance, it has all variants of the globins from all species. It has many serine proteases. Thus we try to remove all homologous genes. Some can be done by mechanically matching sequences, and some is done by hand. We also eliminated internal repeats, such as the collagens or the immunoglobulin superfamily. Having created a database of exons, we then matched those exons against each other for amino acid identity. We did that by comparing exons of roughly similar size and sliding the sequences along each other to count the numbers of exact amino acid matches that there were without gaps. We slid the exons by five residues, to look for cases in which the positions of the introns might have drifted a small amount. This is a simple computer calculation and it gave us a list of amino acid similarities.

Our problem then was to decide when the amino acid matches are significant—when does it stand for something like evolutionary homology, and when is it just some random agreement of the amino acid sequence. We construct such a statistical test by randomizing the database in terms of amino acid sequence and running through the matching program again to ask, how many matches occur on a purely random basis? By running through the database some 20 times, we can get a 95 per cent confidence estimate, because we can demand an amino acid similarity that would not occur more than once on a random basis in 20 simulations. For example, this criterion, for exons of length 90 to 99 amino acids, is that if 20% of the amino acids match, that would occur by chance only once out of 20 times. However, for an exon of length 20 to 29, the similarity has to be better than 45 per cent for the match to be significant.

In the reduced database of 1,255 exons we find 18 examples of exon shuffling, and 19 examples of internal repeats; 18 shuffles suggests that the universe is about 44,000 exons in size. That's a surprising number. A while ago, I pulled the argument out of a hat that perhaps there were a million shapes in the original universe. This is a much smaller number. Furthermore, our estimate is most likely to be an overestimate. The reason is that we demanded a 95 per cent confidence level in our amino acid matching to conclude that a match was due to evolutionary homology. We know that over the course of evolution the degree of amino acid similarity will drift down. The work of x-ray crystallographers has shown that one can have three-dimensional structures that are still identical, and clearly homologous, although the match in amino acid sequence has drifted down to the 10 per cent level; so there must be many more matches than we have found. For that reason, the number of true exon shuffles has to be much larger than 18 for this database, and, therefore, the actual size of the universe from which they were drawn is significantly smaller than 44,000. The universe is much more likely to be on the order of 10,000. This work was done by Rob Dorit and Lloyd Schoenbach in my laboratory.

Not only does the notion of putting the original genes together from small structures alleviate the underlying evolutionary problem of how diverse proteins and organisms were created, but the number of structures that are used to put all proteins together is very, very small. We should soon be able to itemize all of them.

QUESTIONS AND ANSWERS

Q: What's known about any other glycolytic enzymes, for example, pyruvate kinase?

Walter Gilbert: The comparison between plants and animals is not known for that gene. There is a series of enzymes, about a dozen, that have an alpha-beta barrel structure somewhere in the protein. Pyruvate kinase is an example, as is triose phosphate isomerase. So far there is no evidence from comparison of intron/exon structures that the alpha-beta barrels have a common origin. If you believe the intron/exon structure you could argue that these are examples of convergent evolution, in which one has obtained similar three-dimensional structures by putting together pieces using slightly different rules.

Russell Doolittle: Yesterday we heard, in the introduction to this conference, how much we've all learned since the first meeting on evolution was sponsored by The Wistar Institute. It's with that in mind that I really feel bad to see you fall into the same trap that the mathematicians of 1966 were extolling. This business about counting the number of exons that fit on the head of a pin seems to me to be going in a very bad direction because this isn't how evolution works. The major fallacy of the 1966 mathematical approach was that they thought that you had to search all of sequence states. Now admittedly, your trap is a lot shallower than theirs because you've reduced it to something like 20^{10} as opposed to 20^{300} which they were dealing with. But in fact, evolution, as was pointed out yesterday, is tinkerism and opportunity and even giving you original exons, which you know very well I'm not about to, I don't think that that's the way it goes. I think that if it works, it goes, and then after that you are a prisoner of your history. You don't go around trying out ten million things and then saying I like this one better than the others.

Walter Gilbert: I think that's an interesting attitude. However, it's only a polemical attitude. I regard these questions as experimentally accessible, rather than solely matters of principle. It is your belief that evolution is a tinkerer and never takes any more complicated steps. That's a belief, not a fact. The interesting calculation is the one I just showed you. It is the statement that the number of genes we have looked at is now sufficiently large, that we will begin to see examples of exon shuffling. That's not obvious initially. For years actually, going back to 1978, people would say, "There should be lots of examples of exon shuffling. It should be obvious everywhere you look." It's not obvious because the number of genes that we have looked at is still a small fraction of all the genes that are out there. But that fraction is now becoming appreciable. The particular prediction, of course, that's being made here is that if we work out, let us say 10^5 human genes in the human sequencing project, we will discover that those genes are highly redundant, not just that they are broken up into a thousand or so gene families but that they exhaust these 10,000 exon structures. We will find a specific set of relationships between those tenes which we would not predict in terms of your evolution as just a tinkerer mechanism. I do believe—I agree with you on that—that evolution will turn out to have searched a very tiny selection of all the states. Many people don't believe that. Many people believe that evolution in changing protein structures goes through a large fraction of the states. We will find this by experiment, because there are different consequences of the different ways of looking at things.

Q: One thing I've never been certain about, when you talk about the introns existing, say between exon 1 and exon 2 in a plant or animal homologue, and if you were to trace back through the evolutionary lineage to much, much older lineages, do you mean to imply that those intron sequences are really evolutionary-homologous?

Walter Gilbert: The question is really, is there any way of telling whether the introns are at identical positions because they have inserted into positions in some previous existing gene, or are they at identical positions because they are the evolutionary remnants of an original gene structure which had an intron at that position? I believe the latter and I tried to argue, from the triose phosphate isomerase comparison, that it was at least likely to be true over a billion years of evolution. It is not possible to prove one way or the other, over longer lengths of time, whether the introns were there all the way through or have been inserted. There are many examples in gene structures of the loss of introns later in evolution, as we compare genes going down families. In the cases of self-splicing introns, class-I introns that occur in the T-even phages, there are examples in which there may be the addition of introns to previously existing structures. There are models for how introns might be inserted into structures, but those models do not agree with the regularities one sees in exon structure. For example, in all proteins the exons are far more regular in size than one would expect, had introns been inserted into the DNA by some sequence-specific rule at the DNA level. The introns in triose phosphate isomerase, again, are not inserted solely into those regions of the protein which are most variable over evolutionary time. They are also in the most highly conserved regions of protein, essentially as though the rule by which the introns are put in has nothing to do with the question of whether the protein is conserved or not conserved. However, this doesn't answer your question because I think that at the moment there isn't any strong answer. There is clear evidence in late genes that we have homologous genes with introns wildly different in size, by factors of ten, different totally in sequence, in which those introns have a common evolutionary origin and just represent the drift of DNA sequences.

Q: Are parts of the introns conserved?

Walter Gilbert: Yes, in the sense that control sequences of all sorts lie in many of the intron regions. Those are generally short sequences, on the order of 50 or 100 bases at the most, within intron sequences that are 1,000, 10,000, 100,000 bases long. The amount of intron that is not well conserved is much, much larger than that amount that is conserved. The conserved sequences can be found by doing comparisons across species, so my comment about lack of conservation is only true for the bulk of the intron structure. Let me give you an example. The insulin genes in mouse have an intron on the order of a few hundred bases, while the insulin genes in chicken have that same intron, 3,500 bases long. If you look at the myoglobin genes in different species you find wild differences in intron lengths. The most prominent characteristic is that there is no control, no evolutionary selective pressure, that preserves things like intron sequence and intron length.

Q: It occurs to me that you might be able to narrow your search if you could think of some criteria for finding exons that are more likely to be primordial. I'm thinking of looking either at small exons which, it seems to me, ought to be older, since there's nothing in the model to suggest that these exons should shrink, and also, perhaps looking at small proteins, where there's less three-dimensional structure constraint on function and so the change in primary sequence should be less if the original function is conserved. So my question is, have you tried to look at exons either from small proteins or small exons to see if there's a different distribution of exon shuffling?

Walter Gilbert: You're absolutely right. The model that we suggest, that we would love to be able to test, is that we should find a substructure of 20 amino-acid-long exons out of which we'd build more complicated exons. That is hard to test because the only test we have at this point is amino acid similarity. To be significant, 20-amino-acid-long exons have to match at too great a level, 50 per cent or more matched, and almost any evolutionary drift eliminates that level of matching. If we could do a calculation of three-dimensional shape from amino acid sequence, then we could try to match three-dimensional shapes. Unfortunately, one can't do that calculation yet; there's no way of getting an accurate, or even a vaguely good, shape prediction out of amino acid sequence. Even if we could match the shapes, another problem arises, because that begs the question of whether there's a common evolutionary origin to those shapes. So we're better off with an amino acid match where we can at least argue that what we see does represent a common evolutionary origin. The calculation that I just described may change rapidly over time— because the number of exons is going up as the databases increase. By next year, we should have something like a factor of four or more examples. All the numbers will get more interesting, and we should be able to check more subtly how well the statistics work to test the size of the universe by looking not just for pairs of repeats, but for multiple repeats and so forth.

New Perspectives on Evolution, pages 165-173

New Perspectives on Evolution Provided by Protein Sequences

RUSSELL F. DOOLITTLE
Center for Molecular Genetics, University of California, San Diego, La Jolla, California 92093

INTRODUCTION

Nowhere is Darwin's notion of descent with modification more apparent than in the protein sequences that have accumulated in our data bases during the past 30 years. At this point, it is certain that the vast inventory of existing proteins is mostly the result of a continuing expansion by gene duplication. Added to this saltatory diversification has been a large degree of unit shuffling leading to mosaic proteins of mixed ancestry. Fortuitously, the average rate of change for proteins over the course of the last few billion years has not been so fast as to blur the evolutionary trails of either organisms or the proteins themselves. Thus, many proteins common to prokaryotes and eukaryotes have sequences that are so similar that common ancestry cannot be doubted, even though the common ancestor of the two lineages may have existed as long as two billion years ago. Indeed, it is not at all unlikely that the history of most extant proteins will be traceable to a small number of ancestral types. Moreover, it should be possible to correlate the appearance of particular protein types with major adaptations of organisms to their environments. Already we can see the outlines of certain broad protein classes and the lifestyles they accommodate. In this chapter the histories of several protein families are explored briefly, including protein kinases and phosphatases that are parts of various eukaryotic receptor proteins. Also, an extensive G-protein linked receptor family is examined with an eye to what lies before us in the way of reconstructing specific evolutionary events.

RECONSTRUCTING EVOLUTION WITH PROTEIN SEQUENCES

It is widely accepted that all life on Earth today is descended from a common ancestral cellular organism that existed sometime between 1.5 and 3.0 billion years ago. Life itself is thought to have been the natural outcome of a prebiotic stage during which an inventory of essential building blocks accumulated and after which an RNA world materialized (Darnell and Doolittle 1986). How sophisticated that RNA world may have been is still largely unknown, and whether or not its RNA members could have catalyzed the synthesis of more precursor material, for example, remains to be shown. That the wherewithal for cutting and splicing RNA itself existed has been well demonstrated in recent years (Cech and Bass 1986).

At some point a rudimentary protein synthesis scheme evolved in a way that allowed a meaningful correspondence between a polypeptide chain sequence and a nucleic acid sequence (Crick et al. 1976). The code linking existing nucleic acid and protein sequences must trace back to such a start. At that moment in time a genetic expansion was begun that depended on nucleic acid duplication and subsequent modification by errant base pairing. The error rate for RNA replication is much greater than is observed for contemporary DNA synthesis (Holland et al. 1982), mostly because of a lack of editing and mismatch repair. As a result, the opportunities for generating improved gene products were great, and natural selection must have encouraged rapid sequence change during this period.

Sometime during the Earth's first two or three billion years, there was an enormous dislocation whereupon DNA was incorporated as the main genetic material in some organism. Although there may be some relics of earlier lineages extant today, the bulk of the present living world is descended from that first successful RNA to DNA convert. Eventually, the DNA-based lineages split into several major groups, the foremost of which are the prokaryotes, the eukaryotes, and the archaebacteria (Woese 1983). The eukaryotes, more than the other two groups, happened upon a number of systems that have allowed enormously complicated multicellular organisms. Subsequently, a major trifurcation gave rise to the plant, animal and fungal kingdoms. If we just follow our own animal history here, we can see that it was heralded by the invention of cell-cell contacts and signaling devices, one eventual outcome of which was the phylum vertebrata. Some 250 million years ago a mammalian radiation began that culminated in big-brained primates who view the world in stereoscopic color.

The question I pose today is, can we reconstruct some of the details of those major happenings in our history on the basis of contemporary protein sequences? Obviously, it depends in part on how fast the protein sequences have been changing. One could imagine, for example, that the rate of change could have been so rapid that the hemoglobin sequences of mammals and fish might not be recognizably related. Or, for the other extreme, if all protein sequences had changed as slowly as the actins, which can be as much as 90% identical in fungi and animals, then there would not be enough sequence difference to distinguish birds and mammals on this basis.

As it happens, however, different proteins change at quite different rates, and we have at our disposal a battery of good chronometers for judging the past. Better still, there seems to be enough original information remaining in contemporary protein sequences that we can root many, perhaps most, back past the major biological boundary points to the prototype groups themselves (Doolittle 1981). For example, most of the important metabolic pathways were laid down before the divergence of prokaryotes and eukaryotes, and the enzymes that constitute them in the two lineages are, on the average, 40–50% identical (Doolittle et al. 1986). This means that we can use protein sequences to determine when that divergence took place. So far, the data indicate the split occurred 1.8 ± 0.4 billion years ago (Doolittle et al. 1989). This date was determined on the basis of only nine different proteins for which a sufficient number of prokaryotic and eukaryotic sequences had been reported; as more sequences are determined the accuracy should improve. The clock was set in this case by determining how much sequence change has occurred in those organisms for which a fossil record exists, and then extrapolating the change observed between prokaryotic and eukaryotic sequences to find the time of divergence.

Looking back further, we could ask, how many genes and gene products were in existence at the time of the big RNA to DNA conversion? How far back towards that point can we reckon events with present-day sequences? There are some hopeful signs. For instance, it has been realized for a long while that the aminoacyl tRNA synthetases ought to be among the most ancient of protein families (Doolittle 1979). Still, when the first aminoacyl tRNA synthetase sequences were reported, there was so little similarity among them that a solid case for common ancestry could not be made. It wasn't until a dozen or more had been reported that a convincing case for general homology was established (Heck and Hatfield 1988). Even then, not all the known sequences could be rooted to a common source.

But, if we presume the same 20 amino acids in use today were also used by the organisms that successfully switched to DNA, and that the code had already been frozen into place during the RNA world, then it should be possible to fix that historic moment in time and, eventually, to estimate how many protein prototypes existed. This, in turn, should permit a surmise as to how big the genome of that primitive ancestor might have been. Whether we will ever be able to probe back beyond that historical divergence point on the basis of sequence data alone seems doubtful, given the rapid rate of change that must have existed in the RNA world. Still, some vestigial icebergs of resemblances may be discernible.

SOME EXAMPLES OF EUKARYOTIC HISTORY DERIVED FROM PROTEIN SEQUENCES

In my presentation today I am going to concentrate on those relatively recent events that have occurred since the divergence of prokaryote and eukaryotes about two billion years ago. As implied above, many of the relevant sequence relationships are more than sufficiently similar to trace the broad course of events that interests us here. I will particularly emphasize a group of proteins involved in sensing extracellular signals.

But first, let me outline briefly the methodology to be used. What it amounts to is the artful alignment of existing protein sequences in away that best reflects their historical connections. The volume of data is such that we use computers to assist us in the task, a specific set of rules being followed in making the alignment (Feng and Doolittle 1987). The next step is to construct a phylogeny from this sequence alignment. How best to do this is always a matter of debate. The trees I will be discussing today have been generated independently by several quite different approaches, however, including both matrix and character-based procedures. Procedures like these have long been used successfully to study the systematics of organisms (Fitch and Margoliash 1967; Dayhoff and Eck 1969). They have also been useful in charting the evolutionary course of some protein families, beginning with pioneering studies on various hemoglobin chains (Itano 1957; Ingram 1963), and it is this latter application that we will emphasize today. Nonetheless, it is important to remember that in such studies of protein families we depend on species comparisons to set our evolutionary clock.

Shuffled domains. If all we had to concern ourselves with were those events in which entire genes were duplicated, the historical reconstruction would be straightforward. As is well known, however, many duplications involve only parts of genes, and these often correspond to stable folding units encoded by single exons (Gilbert 1985; Doolittle 1985). The resulting mosaic proteins have a mixed ancestry which often confounds simple interpretation.

As an example, consider the ever-expanding immunoglobulin family. The earliest protein sequences available from immunoglobulins revealed that antibody proteins had evolved by a series of tandem gene duplications that had given rise to various heavy and light chains (Singer and Doolittle 1966; Hill et al. 1966). Subsequently, it was discovered that an extracellular protein involved in histocompatibility, β_2 microglobulin, was related to immunoglobulins (Peterson et al. 1972). Since that time, a very large number of proteins, cellular and extracellular alike, have been found to be members of the immunoglobulin family (Williams and Barclay 1988; Hayashida et al. 1988). In many of these cases, the immunoglobulin-like structure exists in combination with other characteristic protein types.

The immunoglobulin basic unit is about 100 amino acids long, and is held in a compact β form by a single disulfide bond. These units tend to pack together in such a way as to form good combining sites for antigenic determinants and other molecules. Given this propensity, it is perhaps not surprising, in retrospect, that we find immunoglobulin-like structures serving as ligand-binders in various receptor molecules (Fig. 1). What is fascinating is that, as a result of shuffled combinations, the responses triggered by some of these

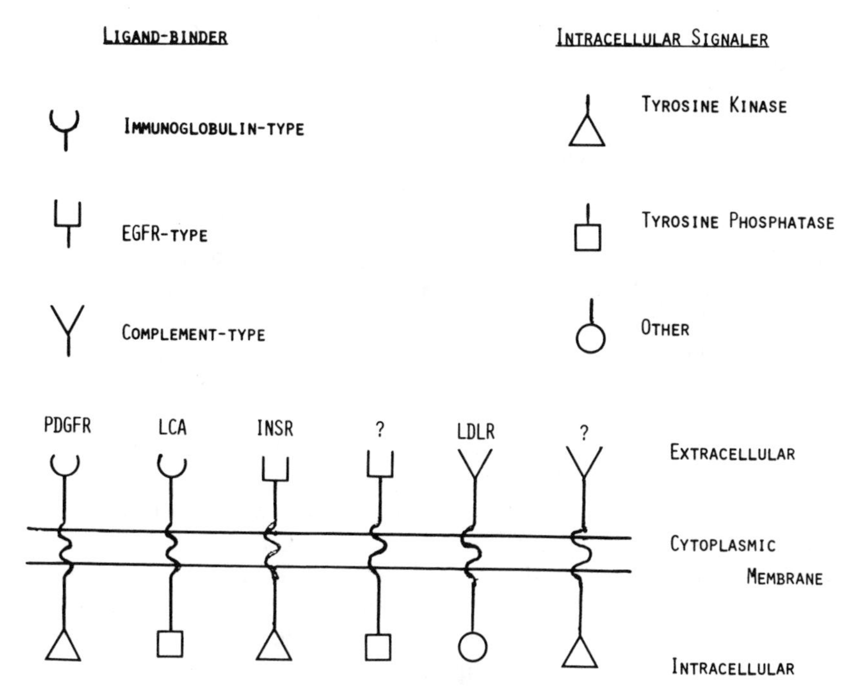

Fig. 1. Eukaryotic receptor proteins have been assembled evolutionarily by shuffling together various combinations of ligand-binding domains with various intracellular response domains. For example, the platelet-derived growth factor (PDGF) receptor has an immunoglobulin-type extracellular unit that binds PDGF, but its intracellular carboxy-terminal portion is a tyrosine kinase. In contrast, the leucocyte common antigen (LCA) also has an immunoglobulin-type extracellular ligand-binding portion, but it has an intracellular tyrosine phosphatase as its intracellular response unit. In all the cases shown the extracellular and intracellular domains are connected by a single membrane-spanning sequence.

different receptors can be exactly opposite. For example, the receptor for platelet-derived growth factor (PDGF) has an extracellular portion that is related to the immunoglobulin family, and its intracellular response domain is a tyrosine kinase. The leucocyte common antigen (LCA), on the other hand, also has an immunoglobulin-like ligand binding domain (Streuli et al. 1988), but its intracellular domain is a tyrosine phosphatase (Charbonneau et al. 1988).

Other combinations of extracellular and intracellular domains exist, and doubtless more will be found (Fig. 1). The question is, will we be able to determine, by sequence analysis alone, when these various concoctions were assembled? It will be difficult, so let us turn to a more tractable set of sequences that appear to have evolved by a straightforward series of gene duplications that seem not to have involved exon shuffling. The gene family I have in mind is a different set of membrane receptor proteins: the so-called G–protein-linked receptors.

G-protein-linked receptors. The G-protein linked receptors are a family of eukaryotic membrane proteins typified by the presence of seven membrane-spanning segments (Kerlavage et al. 1986; Dohlman et al. 1987). Characteristically, these single-chained proteins have an extracellular amino-terminal segment and an intracellular carboxy-terminal segment. What is most interesting about these receptors is the wide variety of ligands and environmental stimuli that can activate them, all the way from the light-stimulated isomerization of retinal in rhodopsin to various pharmacologic stimuli like adrenaline and acetylcholine. The family will almost certainly prove to have hundreds, probably thousands, of members, including the sensory receptors involved in olfaction (Lancet and Pace 1987).

When did all these gene duplications take place? My colleagues and I are attempting to answer that question simply by comparing the protein sequences. The task is daunting, however, in that the sequences are variable in length, ranging from 350 to 600 amino acid residues, and many of the resemblances are quite low-level. Moreover, the multiple membrane-spanning segments bias the amino acid compositions and make it difficult to assess overall differences. Still, we have had some success in that three quite different computer approaches have yielded similar topologies (Fig. 2). Thus, in the most likely phylogeny, a set of fungal receptors clusters as an outgroup, and all the visual pigments are set off in a subgroup by themselves. The vertebrate pharmacologic receptors form a third major cluster with two subclusters (Fig. 2).

We can get some idea of time from this diagram by taking note of the three direct species comparisons that are included. In this regard, the distance back to the divergence of vertebrates and invertebrates can be estimated by the distance back to the divergence of drosophila and bovine rhodopsins (RODM and RHOB, respectively, in Fig. 2); it amounts to about 150 evolutionary units on the scale employed. Similarly, the distance back to the divergence of birds and mammals is obtained from the comparison of mammalian and avian β_2 adrenergic receptors (BARH, HBAR and BARH in Fig. 2); it amounts to about 80 units. The distance back to the primate-rodent divergence (BARH and HBAR) is about 25 units. If we now scale these distances to divergence times obtained from the fossil record, we can get a rough idea of the real times involved in the expansion of the protein family.

For example, assuming equal rates of change, the red-green split of the opsins shown must have occurred in the time since primates and rodents diverged. Pursuing this reasoning, we come to the unavoidable conclusion that much of the basic pharmacology of vertebrates, sympathetic and parasympathetic alike, must be common to invertebrates also. The roots are very deep.

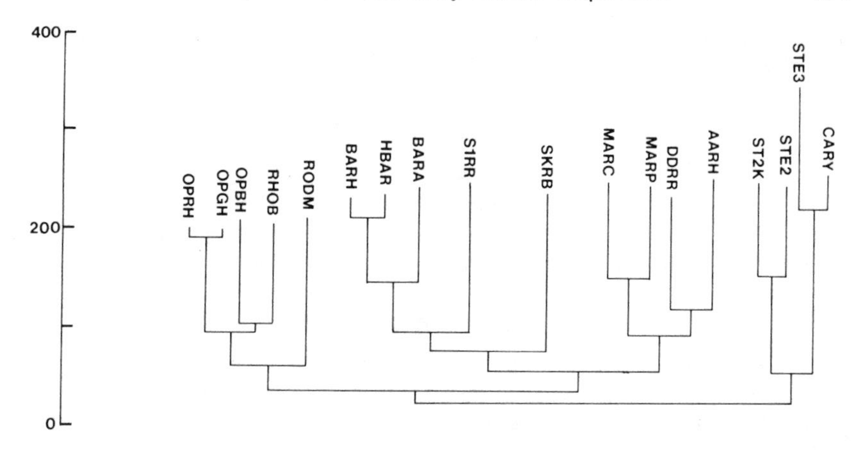

Fig. 2. Phylogenetic relationship of 18 G-protein-linked receptors. This type of receptor is typified by seven membrane-spanning sequences. The extracellular signals sensed are various small molecules and peptides, although in the case of the visual pigments it is the light-induced isomerization of retinal that is the triggering event. OPRH, opsin, red, human; OPGH, opsin, green, human; OPBH, opsin, blue, human; RHOB, rhodopsin, bovine; RODM, rhodopsin, *D. melanogaster;* BARH, beta adrenergic receptor, human; HBAR, beta adrenergic receptor, hamster; BARA, beta adrenergic receptor, turkey; S1RR, serotonin receptor, rat; SKRB, substance K receptor, bovine; MARC, muscarine acetylcholine receptor, cardiac, pig; MARP, muscarine acetylcholine receptor, cerebral, pig; DDRR, dopamine receptor, rat; AARH, alpha adrenergic receptor, human; ST2K, alpha mating factor STE2, *S. cerevisiae;* alpha mating factor STE2K, *S. pombe;* STE3, alpha mating factor STE3, *S. cerevisiae;* CARY, cyclic AMP receptor, *D. discoides.* The distance bar on the left is in evolutionary units.

There is another aspect to this receptor family that needs further scrutiny: the same ligand is involved at two different places on the tree. Thus, adrenaline is the activating agent for both the β_2 adrenergic receptors (BARH, HBAR and BARA in Fig. 2) and the α_2 adrenergic receptor (AARH). Did the two binding sites arise independently, or was the same site shuffled into the two different locations? In this regard, both of these adrenaline receptors are wholly encoded by a single exon, no introns being present. The sequence data are uninformative on the point in that the same phylogenetic tree emerges whether or not the regions thought to encompass the binding sites are included, and no striking resemblances exist locally between the two types.

Eventually, it should be possible to gauge just when in the Earth's history an adrenalin-binding receptor made its first appearance, and the same for the acetylcholine-binding members of this family. In the end, there should be a good correlation with the occurrence of these receptors in the biological world. Thus, if a group of organisms diverged from the mainline before a particular duplication, we naturally expect that particular receptor, and the physiological response it allows, to be absent.

CONCLUSIONS

The availability of vast numbers of amino acid sequences is providing a new perspective on evolution. It should be possible to chart the history of most extant proteins and to correlate the appearance of new protein types with major adaptations of organisms to their environments.

REFERENCES

Cech TR, Bass BL (1986): Biological catalysis by RNA. Annu Rev Biochem 55:599.

Charbonneau H, Tonks NK, Walsh KA, Fischer EH (1988): The leucocyte common antigen (CD45). A putative receptor-linked protein tyrosine phosphatase. Proc Natl Acad Sci USA 85:7182.

Crick FHC, Brenner S, Klug A, Pieczenik G (1976): A speculation on the origin of protein synthesis. Orig Life 7:389.

Darnell JE, Doolittle WF (1986): Speculations on the early course of evolution. Proc Natl Acad Sci USA 83:1271.

Dayhoff MO, Eck RV (1969): "Atlas of Protein Sequence and Structure." Chapter 4. Silver Spring, MD: Natl Biomed Res Found.

Dohlman HG, Caron MG, Lefkowitz RF (1987): A family of receptors coupled to guanine nucleotide regulatory proteins. Biochemistry 26:2657.

Doolittle RF (1981): Similar amino acid sequences: Chance or common ancestry? Science 214:149.

Doolittle RF (1979): Protein evolution. In Neurath H, Hill RL (eds): "The Proteins." New York: Academic Press, p. 1.

Doolittle RF (1985): The genealogy of some recently evolved vertebrate proteins. Trends Biochem Sci 10:233.

Doolittle RF, Anderson KL, Feng DF (1989): Estimating the prokaryote-eukaryote divergence time from protein sequences. In Fernholm, Kare, Jornvall (eds): "The Hierarchy of Life. Molecules and Morphology in Phylogenetic Analysis." Amsterdam: Elsevier Science Publ BV.

Doolittle RF, Feng DF, Johnson MS, McClure MA (1986): Relationships of human protein sequences to those of other organisms. Cold Spring Harbor Symposia on Quantitative Biology 51:447.

Feng DF, Doolittle RF (1987): Progressive sequence alignment as a prerequisite to correct phylogenetic trees. J Mol Evol 25:351 pp 73–85.

Fitch WM, Margoliash E (1967): Construction of phylogenetic trees. Science 15:279.

Gilbert WA (1985): Genes in pieces revisited. Science 228:823.

Hayashida H, Kuma K, Miyata T (1988): Immunoglobulin-like sequences in the extracellular domains of proto-oncogene fms and platelet-derived growth factor receptor. Proc Japan Acad 64:113.

Heck JD, Hatfield GW (1988): Valyl-tRNA synthetase gene of *Escherichia coli* K12 primary structure of the valS gene and homology with other aminoacyl-tRNA synthetases. J Biol Chem 263:857.

Hill RL, Delaney R, Fellows RE, Lebovitz HE (1966): The evolutionary origins of the immunoglobulins. Proc Natl Acad Sci USA 56:1762.

Holland J, Spindler K, Horodyski F, Grabau E, Nichol S, VandePol S (1982): Rapid evolution of RNA genomes. Science 215:1577.

Itano HA (1957): The human hemoglobins: Properties and genetic control. Adv Protein Chem 12:215.

Ingram VM (1963): "The Hemoglobins in Genetics and Evolution." New York: Columbia Univ Press.

Kerlavage AR, Fraser CM, Chung, F-Z, Venter JC (1986): Molecular structure and evolution of adrenergic and cholinergic receptors. Proteins 1:287.

Lancet D, Pace U (1987): The molecular basis of odor recognition. Trends Biochem Sci 12:63.

Peterson PA, Cunningham BA, Berggard J, Edelman GM (1972): β_2 microglobulin—A free Ig domain. Proc Natl Acad Sci USA 69:1697.

Singer SJ, Doolittle RF (1966): Antibody active sites and immunoglobulin molecules. Science 153:13.

Streuli M, Krueger NX, Hall LR, Schlossman SF, Saito H (1988): A new member of the immunoglobulin superfamily that has a cytoplasmic region homologous to the leucocyte common antigen. J Exp Med 168:1523.

Williams AF, Barclay AN (1988): The immunoglobulin superfamily—Domains for cell surface recognition. Annu Rev Immunol 6:381.

Woese CR (1983): The primary lines of descent and the universal ancestor. In Bendall DS (ed): "Evolution From Molecules to Men." Cambridge: Cambridge Univ Press, p. 209.

New Perspectives on Evolution, pages 175-188

The Phylogenetic Significance of Sequence Diversity and Length Variations in Eukaryotic Small Subunit Ribosomal RNA Coding Regions

MITCHELL L. SOGIN
Center for Molecular Evolution, Marine Biological Laboratories, Woods Hole, Falmouth, Massachusetts 02543

INTRODUCTION

A new perspective on the evolution of eukaryotic microorganisms is provided by comparisons of small subunit ribosomal RNA sequences. The extent of rRNA sequence diversity is consistent with eukaryotes representing a distinct evolutionary lineage that may be as old as the eubacterial and archaebacterial lines of descent. The earliest branching eukaryotic groups are represented by diplomonads and microsporidians. The higher kingdoms, including fungi, plants and animals, diverged from each other during a relatively recent period which also gave rise to several other protist lineages. This view of eukaryote evolution challenges interpretations of the fossil record and suggests that our knowledge of eukaryotic biochemical diversity is skewed by a lack of information from protist groups that represent early branchings in the eukaryotic phylogenetic tree.

Evolutionary relationships within the prokaryotic and eukaryotic microbial worlds have been redefined through comparisons of their ribosomal RNA sequences. From partial sequence analyses of small subunit rRNAs (16S-like rRNAs), Woese has demonstrated that prokaryotes comprise two major kingdoms, the Archaebacteria and the Eubacteria (Woese and Fox 1977; Woese 1987). More recently, through comparisons of complete 16S-like rRNA sequences coded by nuclear genomes, our perspective of eukaryotic phylogenetic diversity has been revolutionized (Sogin et al. 1986; Gunderson et al. 1987a; Sogin et al. 1989). Instead of being relatively recent biological inventions derived from some ancestral prokaryotic lineage, eukaryotes appear to be a discrete evolutionary assemblage that may be as old as the archaebacterial and eubacterial lines of descent. In contrast to what appears to have been a relatively recent separation of plants, animals and fungi, protists are represented by a progression of independent branchings, some of which are as ancient as the divergence between the two prokaryotic kingdoms.

Sequence comparisons of rRNAs or their coding regions have gained widespread acceptance among systematic biologists because rRNAs are evolutionary homologous and functionally equivalent in all organisms, their sequence changes sufficiently slowly to allow measurements of even the largest genealogical distances between compared organisms, and they do not undergo transfer between species (if a gene is transferred between species or genera, phylogenies inferred from comparisons of that gene sequence or its RNA or protein product will be that of the transfered gene, not the organisms). The 16S-like rRNAs are particularly well suited for inferring relationships that span both close and distant evolutionary relationships. They contain a statistically significant number of independently variable sites which must be subject to different functional constraints. Regions that display high rates of change are interspersed among domains that are moderately conserved or nearly invariant (Sogin and Gunderson 1987). The conserved elements are distributed throughout the length of the molecule and presumably reflect functional domains that were established in the earliest common ancestors to all living systems. Phylogenetic distances between even the most unrelated of organisms can be estimated from comparisons of the conserved and partially conserved sequences, whereas the high rates of nucleotide substitutions in the non-conserved regions are sometimes valuable for resolving close genealogical relationships (Sogin et al. 1986). Finally, the 16S-like rRNAs are not simple macromolecular lattices to which ribosomal proteins are attached. These rRNAs can be folded into secondary structure conformations which contain as many as fifty helical regions that are are retained in the rRNAs of species separated by vast evolutionary distances. The helical regions represent independent functional domains which can be used to test for convergent evolution. Identical trees inferred from comparisons of sequence domains that can vary independently must reflect divergent evolution from a common ancestor; it is improbable that functionally separate macromolecular sequences will converge at the same rate during evolution and produce identical phylogenetic frameworks.

The isolation and sequence analysis of rRNA coding regions is facilitated by the interspersion pattern of conserved and non-conserved sequence elements. DNA synthesis in primer extension analyses (Sanger and Coulson 1975) of rDNA coding regions can be initiated using oligonucleotides that are complementary to conserved rRNA sequences (Elwood et al. 1985). Such analyses do not even require the identification and isolation of rDNA clones from recombinant genomic libraries. By taking advantage of nearly invariant sequence elements proximal to the 5′ and 3′ regions of 16S-like rRNAs, polymerase chain reaction (PCR) techniques (Mullis and Faloona 1987; Saiki et al. 1988) can be used to rapidly and preferentially synthesize copies of the 16S-like rRNA coding regions (Medlin et al. 1988). Sequences between the conserved elements can be exponentially amplified by repetitive cycles of denaturing duplex DNA, annealing primers complementary to the conserved sequence elements, and then primer extension using DNA polymerase. The products of the primer extension and the original duplex DNA can serve as templates in successive amplification cycles. Within a few hours, several μg of DNA defining 16S-like rRNAs can be obtained from as little as 0.1 nanograms of bulk genomic DNA. The resulting product can be cloned into the single-stranded phage M13 or characterized by modifications of the dideoxynucleotide sequencing protocols for analyzing double-stranded DNA templates. This strategy is rapid, requires a minimal number of organisms for analysis, and—of even greater significance—permits the simultaneous analysis of several representatives from the multi-copy rDNA genes of a single species. By direct sequence analysis of PCR products or concurrent analysis of multiple clones containing representative amplification products, the extent of sequence heterogeneity in the rDNA gene family can be assessed.

PHYLOGENETIC IMPLICATIONS OF SEQUENCE DIVERSITY IN EUKARYOTIC 16S-LIKE RRNAS

Homologous positions in rRNA coding regions must be aligned before the extent of structural similarity between rRNA genes can be determined. Using computer-assisted methods that consider both the conservation of primary and secondary structure, nearly 1000 positions can be unambiguously aligned for rRNAs that represent the three primary lines of descent (eukaryotes, eubacteria, and archaebacteria). The extent of sequence similarity ranges from 68 to 70% between archaebacterial and eubacterial representatives to as low as 55 to 60% between eukaryotic and archaebacterial or eukaryotic and eubacterial rRNAs. Within the eukaryotic lineage the extent of sequence variation exceeds that seen in the entire prokaryotic world. For example, *Giardia lamblia* versus *Plasmodium berghei,* or *G. lamblia* versus *Trypanosoma brucei,* structural similarity values are 0.675 and 0.677, respectively. This compares to values of 0.700 and 0.711 for *Sulfolobus solfataricus* versus *Escherichia coli,* or *Halobacterium volcanii* versus *E. coli,* respectively (Sogin et al. 1989).

The unexpected sequence variation in the eukaryotic lineage is depicted in the phylogeny shown in Figure 1. For this analysis, similarity values between the rRNA sequences were converted to evolutionary distances as previously described (Elwood et al. 1985) and used to infer a phylogenetic tree by the distance matrix methods (Fitch and Margoliash 1967). Divergent representatives for the eubacterial, archaebacterial and eukaryotic kingdoms are included in this analysis. Evolutionary distances are proportional to line segment lengths separating organisms and nodes. For each of the primary lines of descent the branching order is as shown, but the root of the entire tree is not indicated. Within the eukaryotic lineage it can be seen that the earliest diverg-

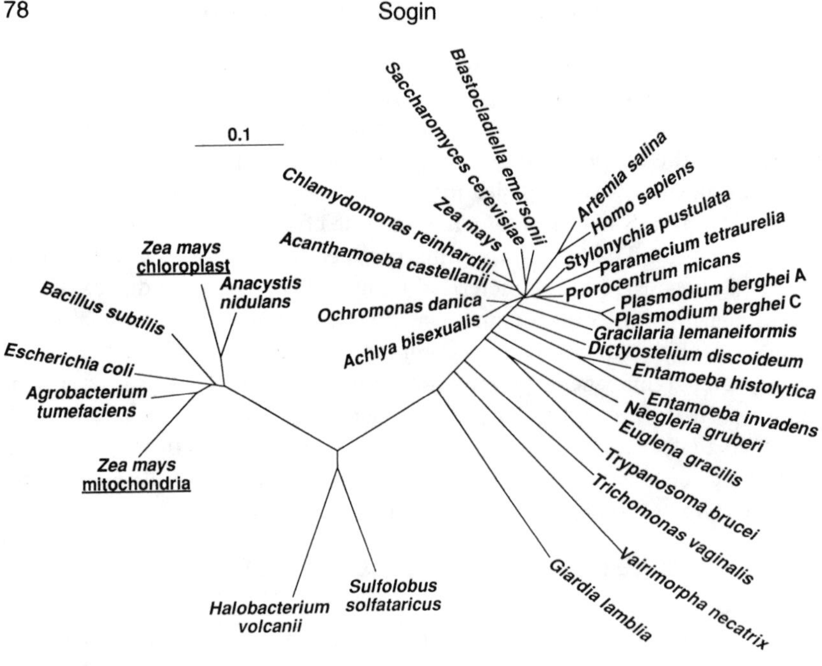

Fig. 1. Multi-kingdom tree inferred from 16S-like rRNAs A computer-assisted method was used to align the 16S-like rRNA sequences from divergent representatives of the Eubacteria, Archaebacteria and Eukaryota. The alignments were influenced by considering the evolutionary conservation of both primary and secondary structure features (Elwood et al. 1985). The distance matrix methods (Fitch and Margoliash 1967) were used to infer an unrooted multi-kingdom tree in which the line segment lengths represent the evolutionary distance between organisms.

ing lineage is that represented by *G. lamblia,* and the evolutionary distances within the eukaryotic subtree exceeds similar measurements within the entire prokaryotic world. Based upon rRNA sequence diversity it would appear that eukaryotes represent an ancient lineage as old as the eubacterial and archaebacterial lines of descent.

The topology and depth of divergence within the eukaryotic subtree may be influenced by the high G/C content in the 16S-like rRNAs of *G. lamblia* (75% G/C) and *S. solfataricus* (67% G/C) (24). This could lead to convergence towards G or C at many sites, which might influence the position of *G. lamblia* in the tree. Since rigorous methods of compensating for biased G/C content have not been developed, two different approaches were employed to evaluate the phylogenetic position of *G. lamblia.* In the first, distance matrix trees were constructed which only included prokaryotes with low G/C contents in their rRNAs or NWS alignment scores (Smith et al. 1981) were used as estimates of evolutionary distances. In both cases, *G. lamblia* still represented the deepest branching lineage in the eukaryotic line of descent.

In a second analysis, nucleotide usage patterns were identified in absolute consensus sequences for a collection of six divergent archaebacteria, six diverse eubacteria, or a composite collection of these archaebacterial plus eubacterial 16S-like rRNAs (Sogin et al. 1989). These usage patterns were compared to typical eukaryotic rRNAs including that of *G. lamblia.* (Positions that were invariant in the alignment of rRNAs from the three primary lines of descent do not contribute evolutionary information and were therefore excluded from the analysis.) Table I displays the number of sites that are identical in comparisons of representative eukaryotic 16S-like rRNAs with archaebacterial or eubacterial modified consensus sequences. These sites are enumerated in four categories: Archaebacteria, Eubacteria, Archaebacteria or Eubacteria, and Archaebacteria plus Eubacteria (prokaryote). In most eukaryotic 16S-like rRNAs the nucleotide compositions at 130–170 sites are identical to the archaebacterial (Table I, column 1) or eubacterial (Table I, column 2) modified consensus sequences. Similarly, the "or" comparisons (Table I, column 3)

TABLE I. Absolute Consensus Sequences for 16S-Like rRNAs From Archaebacteria, Eubacteria, or Prokaryotes (Archaebacteria Plus Eubacteria)

	Number of sites identical to positions modified 16S rRNA consensus sequences from			
	Archaebacteria	Eubacteria	Archaebacteria or eubacteria	Archaebacteria plus eubacteria
Organism				
Homo sapiens	162	169	170	103
Xenopus laevis	140	145	186	99
Saccharomyces cerevisiae	146	144	194	96
Zea mays	140	143	189	94
Dictyostelium discoideum	140	146	191	96
Euglena gracilis	144	142	194	93
Trypanosoma brucei	130	135	180	86
Vairimorpha necatrix	106	103	146	63
Giardia lamblia	296	248	395	149

Universally conserved positions (invariant sites within the archaebacterial, eubacterial, and eukaryotic 16S-like rRNAa employed in this analysis) were deleted from the consensus and these modified consensus sequences were compared to representative eukaryotic 16S-like rRNAs. Eubacterial and archaebacterial 16S-like rRNAs represented in the analysis include *Agrobacterium tumefaciens, Anacystis nidulans, Bacillus subtilis, Escherichia coli, Mycoplasma capricolum, Proteus vulgaris, Pseudomonas testosteroni., Halobacterium cutirubrum, Halococcus morrhuae, Halobacterium volcanii, Methanococcus vannielii,* and *Sulfolobus solfataricus* (Dams et al. 1988), and *Pyrodictium occultum* (Woese, unpublished data).

show that 170–194 positions are identical to either the archaebacterial or the eubacterial modified consensus sequences. All values for *Vairimorpha necatrix* are unusually low; this probably reflects the accelerated rate of evolutionary change in the *V. necatrix* 16S-like rRNA and its relatively short length of 1250 nucleotides (Vossbrinck et al. 1987). Conversely, the values for *G. lamblia* are unusually high. The *G. lamblia* 16S-like rRNA coincides with the archaebacterial and eubacterial modified consensus sequences at 296 and 248 sites, respectively. This is reflected in the prokaryotic modified consensus sequence comparison where 149 positions are identical to the *G. lamblia* rRNA (Table I, column 4). The *G. lamblia* sequence has 395 positions that are either eubacterial or archaebacterial in character (Table I, column 3). These 395 positions are not localized within a few helices that might have been imported by transposition events into the *G. lamblia* rRNA coding regions. Instead they are distributed throughout a proposed secondary structure as shown in Figure 2. *G. lamblia* appears to have retained many of the features present in archaebacterial or eubacterial 16S-like rRNAs including a putative Shine-Delgarno sequence. (The 3′ proximal Shine-Delgarno sequence in prokaryotic 16S-like rRNAs is complementary to elements conserved in prokaryotic mRNAs. Analogous regions in eukaryotic 16S-like rRNAs complementary to eukaryotic mRNAs have yet to be reported.) We infer that these features were also present in ancestral 16S-like rRNA sequences common to the archaebacterial, eubacterial and eukaryotic lines of descent. This signature analysis is consistent with our distance matrix tree, which shows *G. lamblia* as the earliest diverging lineage in the eukaryotic subtree.

Phylogenetic trees similar to that presented in Figure 1 should be regarded as frameworks against which the utility of of alternative evolutionary markers or traits can be evaluated. For *G. lamblia,* its early branching is supported by a lack of mitochondria, the apparent absence of ER, the evident lack of sexual life cycle stages (Feely et al. 1984), and the remarkably simple constellation of proteins associated with its cytoskeleton (D. Peattie, unpublished results). *G. lamblia* probably separated from other eukaryotes after the appearance of a flagellar apparatus but prior to the full development of the ER, earlier than the endosymbiotic event(s) that gave rise to mitochondria, and before the cytoskeleton had reached the level of complexity found in other eukaryotic microorganisms.

Soon after the divergence of *G. lamblia,* several early branchings occurred including the microsporidian V. necatrix and the trichomonad *Trichomonas vaginallis.* All three are protozoans which have adopted parasitic life styles and lack mitochondria. The parasitic life style may have played a critical role leading to the survivorship of these lineages during unfavorable periods in their evolutionary history. However, parasitism alone can not explain early branching lineages; free living diplomonads related to *G. lamblia* can be identified in

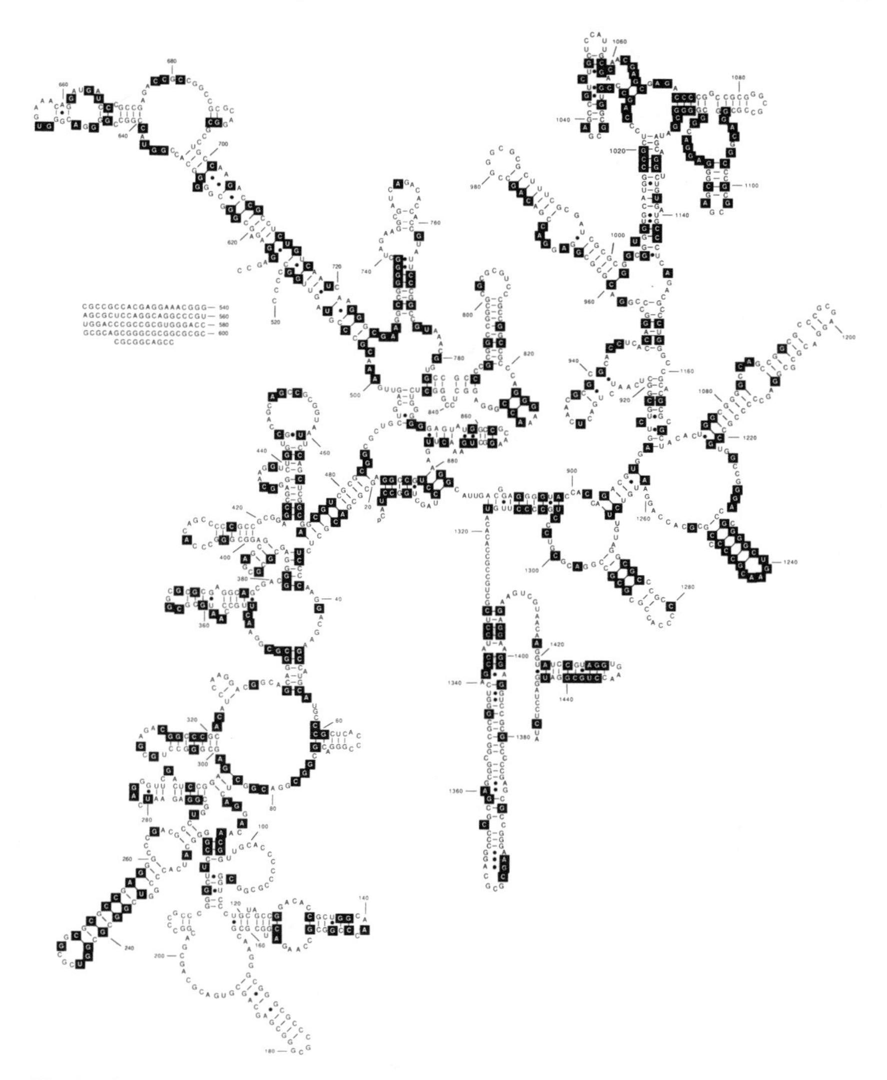

Fig. 2. Secondary structure for the Giardia lamblia 16S-like rRNA. Positions in the *G. lamblia* 16S-like rRNA that are identical to modified consensus positions in either the archaebacteria or eubacteria (see Table I, column 3) are indicated by reversed letters. The helical regions are based upon their phylogenetic conservation (Noller and Woese 1981).

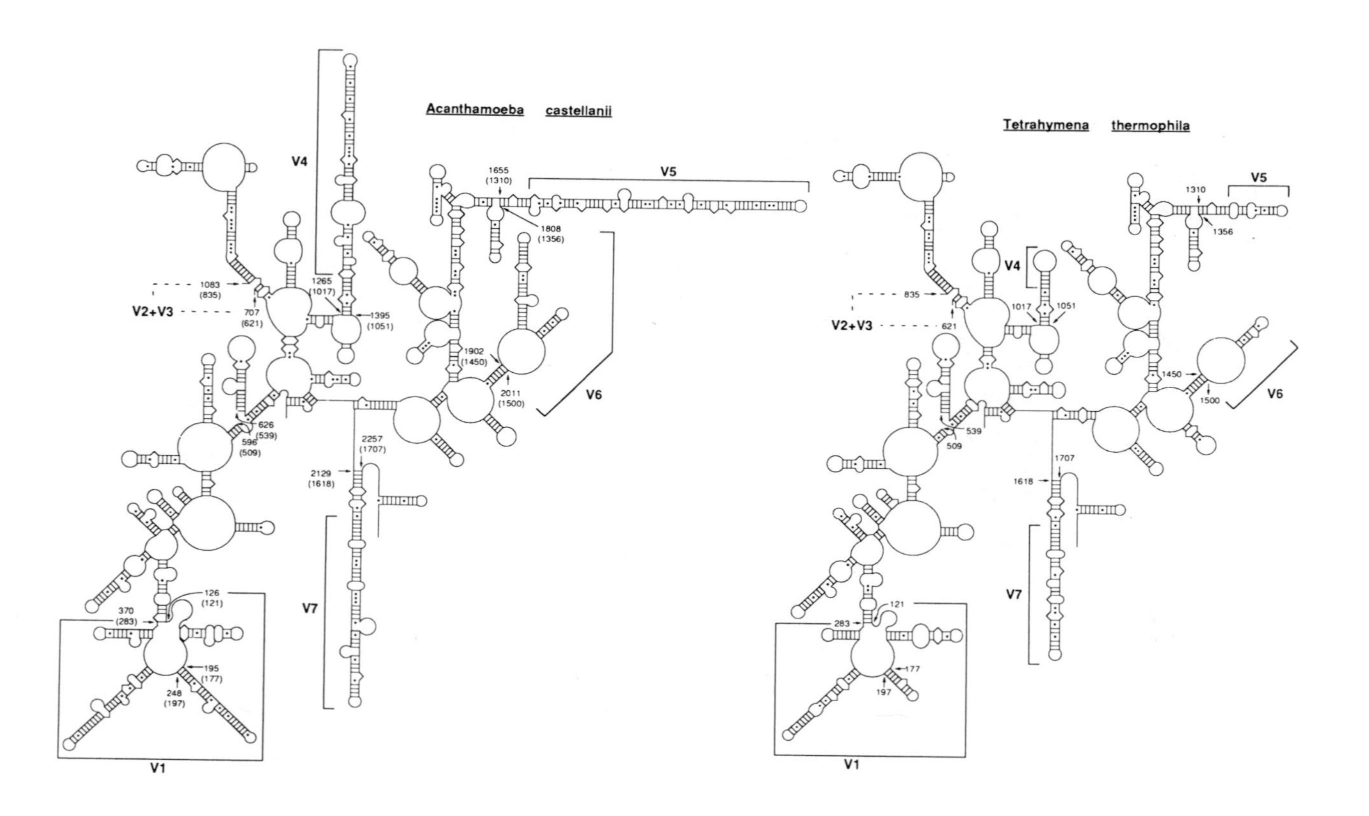

Acanthamoeba castellanii
Tetrahymena thermophila
V1
V2+V3
V4
V5
V6
V7

our contemporary biosphere and other parasitic species including *Plasmodium* and *Pneumocystis* display late branching patterns.

The early branchings are followed by a collage of seemingly unrelated protist lineages. This contrasts with the plants, animals, and fungi which appear as monophyletic groups. These higher kingdoms separated nearly simultaneously late in the evolutionary history of the nucleated cells during a period which also gave rise to numerous other protist groups including chlorophytes/plants, acanthamoebae, ciliates plus dinoflagellates, rhodophytes, and a morphologically diverse group containing chrysophytes, diatoms, brown algae, and oomycetes. Branchings for these major lineages span a remarkably short time-frame. This precludes the identification of the proper order of succession for the nearly concurrent diverging lineages. Their nodes are separated by fewer than one nucleotide change per one hundred positions. Yet, the general branching pattern shown in Figure 1 is nearly constant regardless of which eukaryotes are included in the analysis.

LENGTH VARIATIONS IN EUKARYOTIC 16S-LIKE RRNAS

In addition to sequence diversity, considerable length variations are observed in eukaryotic rRNA coding regions. *V. necatrix* rRNA contains 1250 nucleotides (Vossbrinck et al. 1987) while the *Acanthamoeba castellanii* mature rRNA transcript contains as many as 2305 residues (Gunderson and Sogin 1986). There is no correspondence between length of a given eukaryotic 16S-like rRNA and its position in the rRNA based phylogenetic trees. The length variations are primarily confined to seven highly diverse sequence domains in mature 16S-like rRNA transcripts. Examples of length variations in eukaryotic rRNA genes are indicated by the expansion regions V1–V7 in the *T. thermophila* and *A. castellanii* 16S-like rRNA secondary structures, presented in Figure 3. In general, regions which display extreme length variation can be folded into long helical structures but these are not considered to be phylogenetically proven using the criteria of compensatory base changes (Noller and Woese 1981). The lower rates of evolutionary change in these regions relative to noncoding genomic DNA (Larson and Wilson 1989) may reflect a requirement to maintain the highly variable regions in higher order structures that do not interfere with the functional roles of the 16S-like rRNA.

Fig. 3. Location of hypervariable regions in eukaryotic 16S-like ribosomal RNAs. The secondary structures for the Tetrahymena thermophila and Acanthamoeba castellanii 16S-like rRNAs were drawn according to the pairing schemes previously proposed for eukaryotic small subunit rRNAs (Gutell et al. 1986). The brackets V1–V7 denote regions that display extreme length variation. The numbers indicate absolute sequence positions in each 16S-like rRNA and the numbers in parentheses for the *A. castellanii* structure denote the corresponding positions in the *T. thermophila* 16S-like rRNA.

It is not known if extra nucleotides are introduced into expansion regions of rRNA genes by transposition events or if the variable regions grow by multiple insertions of small numbers of nucleotides. Indirect evidence supporting the transposition theory comes from the identification of extra nucleotide stretches which interrupt highly conserved sequences in several large subunit (23S-like) rRNA coding regions. In all known examples, extra nucleotides found in highly conserved regions of rRNA coding regions are removed from the primary transcripts by post-transcriptional RNA processing mechanisms. All are members of the group-I introns and in *Tetrahymena* they are capable of self-splicing (Kruger et al. 1982). We have previously argued that *Tetrahymena* rRNA introns were acquired subsequent to the evolutionary divergence of different *Tetrahymena* species (Sogin et al. 1986). This conclusion is based upon the incongruent phylogenies inferred from comparisons of the intron sequences with phylogenetic frameworks based upon similarities between their 16S-like rRNA genes or comparative analyses of alternative evolutionary markers including cytoskeletal proteins, restriction maps for ribosomal RNA genes, and electrophoretic behavior of many enzymes. More recently we have identified similar group-I introns in the 3′ proximal helical region of the *Pneumocystis carinii* 16S-like rRNA (Sogin and Edman, 1989). The similarity of the *P. carinii* rRNA intron to those of *Tetrahymena* 23S-like rRNAs is most likely due to their divergence from a common ancestral sequence. An alternative, but less plausible possibility, is that these introns are products of convergent evolution. If the *Tetrahymena* and *P. carinii* introns share a common evolutionary history, their occurrence in rRNA genes of unrelated species and nonhomologous coding regions must be a consequence of lateral gene transfer events.

Regions that are highly conserved in rRNAs must serve pivotal functions in the ribosome. The introduction of extra nucleotide sequences into highly conserved domains in rRNA coding regions would effectively destroy the function of the rRNA gene unless the extra nucleotide elements can be efficiently removed by appropriate RNA processing machinery. Once the intron has transposed into a functionally important domain in the rRNA, it becomes subject to selective pressure which insures its accurate and efficient excision from primary transcripts. Failure to meet this requirement would effectively destroy the function of the rRNA gene and probably lead to it becoming a pseudo-gene.

In contrast, expansion sequences in rRNAs are poorly conserved and presumably do not play critical functional roles. It is conceivable that they also resulted from transposition events but are usually not subject to selective pressures similar to those imposed upon the group-I introns that reside within highly conserved regions of rRNA genes. If the hypervariable regions have grown by transposition events similar to those which incorporated group-I introns into highly conserved rRNA coding regions, it is unlikely that the imported elements will have retained the sequence character of the original transposon. However, under certain circumstances hypervariable regions can assume functional roles in the ribosome. *Plasmodium berghei* exemplifies this possibility. In this organism, two different types of rDNA transcription units have been described with two copies of each gene being dispersed throughout the genome (Dame et al. 1984). The two classes define similar ribosomal RNAs but the internal transcribed spacer regions, and the flanking sequences, display considerable sequence variation. Heteroduplex analyses have shown that only one of the rDNA gene classes, the "A" gene, is colinear with ribosomal RNA species isolated from asexual bloodstream parasites. Similar experiments demonstrated that the second gene, referred to as the "C" gene, is not colinear with mature rRNA transcripts isolated from *P. berghei* grown in the mammalian host. Sequence analysis of both genes revealed that the "A" gene is 96.5% homologous with the "C" gene (Gunderson et al. 1987b). Positions which vary between the two sequences are not randomly dispersed throughout the length of the molecule. The non-random distribution of nucleotide differences between the "A" and "C" genes is an indication that the inactive "C" gene may code for a functional rRNA transcript. If the "C" coding region were a pseudo-gene, differences between the two rDNA gene classes should not be concentrated in a limited number of regions. Furthermore all the primary and secondary structural features which are conserved in typical eukaryotic 16S-like rRNA coding regions are also present in the "C" gene. Despite the absence of "C" gene transcripts in *P. berghei* growing in the mammalian host, the "C" gene is still under selective pressure and does not define a pseudo-gene. Further analysis of transcripts by hybridization and primer extension analyses demonstrated that the "C" gene is expressed in *P. berghei* isolated from salivary glands of mosquitos where the organism forms sporozoites. Similar analyses demonstrated that "A" gene transcripts are either repressed or have a short half-life when *P. berghei* enters the sexual stage of its life cycle. This suggests that a major structural modification of the ribosome in *P. berghei* occurs during different stages of its normal life cycle. A coarse control over protein synthesis during the organism's transition between developmental stages could be imagined in which major structural alterations of ribosomes mediate the selection of specific families of mRNAs to be translated; however, evidence to support this possibility has yet to be presented.

CONCLUSIONS

The sequence diversity and length variations in eukaryotic 16S-like rRNAs pose challenging questions to both molecular and organismal biologists. The extreme depths of branching in the eukaryotic subtree are equivalent to or are greater than the depths of branching observed within the entire prokaryotic world. From eukaryotic rRNA sequence diversity it is reasonable to infer that the divergence between the eukaryote and prokaryote lineages occurred early in the evolutionary history of this biosphere. The alternative explanation for the unexpected evolutionary diversity of eukaryotes is that they represent a rapidly evolving lineage which diverged from some ancestral prokaryotic lineage well after the separation of the Archaebacteria from the Eubacteria. To date, the analysis of rRNA sequence data has not convincingly identified a prokaryotic lineage specifically ancestral to all eukaryotic nuclear ribosomal RNA genes. This result conflicts with most interpretations of the fossil record. The origins of prokaryotic microorganisms can be placed at least as early as 3.5 billion years ago (Schopf and Walter 1983) but the earliest eukaryotic microfossils do not appear until the Proterozoic, and hence, the origins of nucleated cells are often placed within the last 1 to 2 billion years. The absence of eukaryotic microbial fossils older than two billion years and the apparent lack of significant biochemical diversity in eukaryotic systems are frequently cited as evidence for a recent origin for the eukaryotic lineage. Such interpretations may have distorted our view of microbial evolution. The identification of eukaryote microfossils is generally based upon their minimal size approaching 10 microns. Since there are extant eukaryotes as small as 1 micron in diameter, e.g., *Nanochlorum eukaryotum,* and since soft-bodied protozoans such as diplomonads are unlikely to be preserved in the fossil record, the lack of fossil record support for extremely ancient eukaryotes is not surprising. Similarly, assessments of eukaryotic biochemical diversity may be misleading. Most information about pathways and biochemical behavior is derived from studies of animals, plants or single cell organisms such as *Saccharomyces.* These groups represent recent branchings in the eukaryotic line of descent, which is consistent with their having similar biochemical motifs. The true biochemical diversity may be represented by euglenoids, kinetoplastids, microsporidians and diplomonads; the rRNA tree suggests that these organisms represent the earliest branching lineages in the eukaryotic line of descent.

A second major question posed by the rRNA sequence analyses concerns the reason for the massive radiation of morphologically divergent forms. This radiation might reflect major changes in the environment that occurred approximately 800 to 1200 million years ago. A large increase in oxygen could have lead to the development of new ecological niches. A similar scenario would be the rapid diversification of a small number of lineages that survived some cataclysmic event. It is also conceivable that the radiative period of evolution was not triggered by a major environmental perturbation. The rate of phenotypic evolution may have suddenly accelerated in response to novel mechanisms related to the management of genetic information. For example, control elements similar to homeo boxes required for complex developmental pathways in multicellular organisms might have evolved. Another intriguing possibility is that "cis" splicing mechanisms for processing RNA might have been invented or vectors for rapidly exchanging genetic information between species may have appeared. Both would facilitate shuffling and exchange of genetic elements which could have led to dramatic phenotypic effects.

The shuffling of genetic information via lateral gene transfer mechanisms may have played a crucial role in the evolution of nucleated cells. Even for rDNA genes which are not capable of undergoing lateral gene transfer ensemble, it is apparent that certain genetic elements are capable of invading both the highly conserved and non-conserved sequence elements. In the former case, the invasion is only tolerated if the element can be excised from primary transcripts. Comparisons of the intervening sequences must not be used to infer organismal phylogenies. In contrast, it is possible and even likely that non-conserved sequences can accommodate transposed genetic information. In some case such invasions might lead to new or altered functions in the ribosome. However, the inference of relationships from comparisons of hypervariable regions will be misleading if the sequences are not evolutionary homologues or if they were brought into the rRNA coding regions by transposition. When relationships are inferred from comparisons of these regions it will be necessary to determine if the topologies of independent hypervariable regions are congruent with each other.

There is no evidence to suggest that moderately or highly conserved elements in rDNA genes are capable of undergoing lateral gene transfer. It is conceivable that these genetic elements are unusual in this regard. They define critical functional domains in a complex ribonucleoprotein particle which must interact at some time with every other protein in the cell. These genetic elements define highly coordinated and complex functions and they can not tolerate sudden, unilateral change. Unlike many other genes which define structural proteins, the conserved elements in rRNAs provide the documents of evolutionary change (Zukerkandl and Pauling 1965). Other gene sequences with evolutionary properties similar to those of rRNA genes may some day be identified but the potential role of lateral gene transfer mechanisms in the evolution of the eukaryotic cell is still not understood.

ACKNOWLEDGMENTS

This work was supported by grants from the National Institutes of Health Grant (GM32964) and the John D. and Catherine T. MacArthur Foundation.

REFERENCES

Dame JB, Sullivan M, McCutchan TF (1984): Two major sequence classes of ribosomal RNA genes in *Plasmodium berghei.* Nucl Acids Res 12:5943.

Edman JC, Sogin ML (1989): A self-splicing intron in the small subunit rRNA gene of *Pneumocystis carinii* Nucl Acids Res 17:5349.

Elwood HJ, Olsen GJ, Sogin ML (1985): The small subunit ribosomal RNA gene sequences from the Hypotrichous ciliates *Oxytricha nova* and *Stylonychia pustulata.* Mol Biol Evol 2:399.

Feely DE, Erlandsen SL, Chase DG (1984): "*Giardia* and *Giardiasis.*" New York: Plenum Press.

Fitch WM, Margoliash E (1967): Construction of phylogenetic trees. Science 155:279.

Gunderson JH, Sogin ML (1986): Length variations in eukaryotic rRNAs: small subunit rRNAs from the protists *Acanthamoeba castellanii* and *Euglena gracilis.* Gene 44:63.

Gunderson JH, Elwood HJ, Ingold A, Kindle K, Sogin ML (1987a): Phylogenetic relationships between chlorophytes, chrysophytes and oomycetes. Proc Natl Acad Sci USA 84:5823.

Gunderson JH, Sogin ML, Wollett G, Hollingdale M, de la Cruz VF, McCutchan TF (1987b): Structurally distinct, stage specific ribosomes occur in *Plasmodium.* Science 238:933.

Gutell RR, Weiser B, Woese CR, Noller HF (1986): Comparative anatomy of 16S-like ribosomal RNA. Prog Nucleic Acid Res Mol Biol 32:155.

Kruger K, Grabowski PJ, Zaug AJ, Sands J, Gottschling DE, Cech TR (1982): Self-splicing RNA: Autoexcision and autocyclization of the ribosomal RNA intervening sequence of *Tetrahymena.* Cell 31:147

Larson A, Wilson AC (1989): Patterns of ribosomal RNA evolution in salamanders. Mol Biol Evol 6:131.

Medlin L, Elwood HJ, Stickel S, Sogin ML (1988): Sequence analysis of enzymatically amplified genomic small subunit rRNA genes from the diatom, *Skeletonema pustulata.* Gene 71:491.

Mullis KB, Faloona FA (1987): Specific synthesis of DNA in vitro via a polymerase-catalyzed chain reaction. Methods in Enzymology 155:335–350.

Noller HR, Woese CR (1981): Secondary structure of 16S ribosomal RNA. Science 212:403.

Saiki R, Gelfand DH, Stoffel S, Scharf SJ, Higuchi R, Horn GT, Mullis KB, Erlich HA (1988): Primer-directed enzymatic amplification of DNA with a thermostable DNA polymerase. Science 239:487.

Sanger F, Coulson AR (1975): A rapid method for determining sequences in DNA by primed synthesis with DNA polymerase. J Mol Biol 94:441.

Schopf JW, Walter MR (1983): In Schopf JW (ed), "Earth's Earliest Biosphere: Its Origin and Evolution," Princeton: Princeton Univ Press, pp 214–239.

Smith TF, Waterman MS, Fitch WM (1981): Comparative biosequence metrics. J Mol Evol 18:38.

Sogin ML, Gunderson JH (1987): Structural diversity of eukaryotic small subunit ribosomal RNAs: Evolutionary implications. In "Endocytobiology III." Annals of the NY Acad Sci 503:125.

Sogin ML, Ingold A, Karlok M, Nielsen H, Engberg J (1986): Phylogenetic evidence for the acquisition of ribosomal RNA introns subsequent to the divergence of some of the major *Tetrahymena* groups. EMBO 5:3625.

Sogin ML, Elwood HJ, Gunderson JH (1986): Evolutionary diversity of the eukaryotic small subunit rRNA genes. Proc Natl Acad Sci USA 83:1383.

Sogin, ML, Gunderson JH, Elwood HJ, Alonso RA, Peattie DA (1989): Phylogenetic significance of the kingdom concept: An unusual eukaryotic 16S-like ribosomal RNA from *Giardia lamblia.* Science 243:75.

Vossbrinck CR, Maddox JV, Friedman S, Debrunner-Vossbrinck BA, Woese CR (1987): Ribosomal RNA sequence suggests microsporidia are extremely ancient eukaryotes. Nature 362:411.

Woese CR (1987): Bacterial evolution. Microbiol Rev 51:221.

Woese CR, Fox GE (1977): Phylogenetic structure of the prokaryotic domain: The primary kingdoms. Proc Natl Acad Sci USA 74:5088.

Zuckerkandl E, Pauling L (1965): Molecules as documents of evolutionary history. J Theor Biol 8:357–366.

New Perspectives on Evolution, pages 189-207

Implications of Radical Evolutionary Changes in Early Development for Concepts of Developmental Constraint

RUDOLF A. RAFF, GREGORY A. WRAY, AND JONATHAN J. HENRY
Institute for Molecular and Cellular Biology and Department of Biology, Indiana University, Bloomington, Indiana 47405

INTRODUCTION

Ideas on the role of developmental processes in governing the course and direction of evolutionary changes in morphology have been predicated largely on theoretical assumptions about the nature of development. A major prediction is that constraints inherent in developmental processes per se limit evolutionary possibilities. Although this concept can be applied to any stage of development, early development is often held to be particularly subject to constraint, and thus to be highly conserved in evolution. Yet early development does evolve, and sometimes dramatically. We have studied two congeneric sea urchins that differ radically in early development. Profound changes have taken place in fundamental processes of early development such as cell fate determination, cell division patterns, timing of cell lineage events, and patterns of gene expression. These experimental studies of radical changes in sea urchin embryos suggest that hypotheses of developmental constraint applied to early development are often exaggerated and misleading. We suggest that somewhat later stages of development, when cellular interactions are at their maximum in establishing global aspects of body morphogenesis, are probably more subject to constraints. The final stages of ontogeny may again be less subject to constraints because development of individual body parts is more locally governed.

THE CONCEPT OF CONSTRAINTS IN EARLY DEVELOPMENT

Although development is still a weakly integrated part of evolutionary biology, the partners are willing. By now there is no question that changes in body structure that we see in the fossil record had their origins in underlying evolutionary modifications of development. Clearly, we must gain a better understanding of how developmental processes evolve. The connection is more profound than simply an accumulation of changes as successive ontogenies track the demands of natural selection. Modes of development themselves influence aspects of life history, choice of adaptive zone, competition, and susceptibility to extinction (Gould 1977, Jablonski 1986, Mileikovski 1971, Strathmann 1985). Evolution of developmental processes mechanistically links evolving genotype with novel phenotype and, very possibly, existing developmental processes may constrain and channel evolution (Alberch 1982, Maynard-Smith et al. 1985).

The concept of developmental constraint has two major roots. The first lies in the observations of paleontologists that many evolutionary histories appear to exhibit long-term trends (Eldredge 1985). It seems unlikely that low levels of directional selection should exist for very long times to guide such trends as, for example, the evolution of the mammalian jaw over 100 million years. It also is evident from the fossil record, as well as from existing diversity, that not all possible morphologies exist for any given body plan (Alberch 1982). This generalization is well illustrated by the apparent limitations to evolutionary modification of digit patterns in the tetrapod limb (Holder 1983) and in other aspects of limb evolution (Oster et al. 1988). Similarly, the 500 million-year history of evolution of the molluscan shell has filled only a limited portion of possible shell morphologies (Raup 1966). At one extreme these phenomena might be due solely to natural selection: missing forms may simply not be selectively advantageous. On the other hand, developmental systems may not be free to continuously vary with selection (Alberch 1982). Thus, the limited domains actually observed might instead reflect basic limitations to evolutionary modification of developmental programs.

The second source of the concept of developmental constraint lies within the history of embryology. In 1828 von Baer published his famous laws of development. These empirical rules state that within any group of animals (in the case of his work, vertebrates), general features appear earlier in development than specific ones; that general features give rise to specific features; that as embryos of related forms develop they diverge from each other; and that embryos of a higher form never resemble the adults of their relatives, only their embryos. In the late 19th century, Ernst Haeckel recast these laws into an evolutionary form, and made them the basis of his principle that ontogeny recapitulates phylogeny. He envisaged that evolutionary changes were introduced via addition of new stages of development to the end of an existing ontogeny (reviewed by Gould 1977). This scheme requires that the ancestral ontogeny be progressively compressed in length; the development of an animal thus passes through (recapitulates) the entire phylogenetic history of its ancestors. Haeckel, of course, recognized that there were interpolations and adaptations to the needs of development at various stages; to hold otherwise would have been to reduce the concept of recapitulation to absurdity. However, Haeckel regarded the study of recapitulation as the best way to trace evolutionary histories, and considered interpolations into ontogenies as inconveniences to be sorted out and put aside.

The strictly Haeckelian view has been long derided (Sedgwick 1894, Garstang 1922, deBeer 1958, Gould 1977), yet its influence is still strong. This is partly because recapitulation does occur in some instances. The use of developmental features has been very useful in phylogeny at the levels of phyla and classes, because many aspects of development are conservative and in many cases embryos do provide links between groups: echinoderms and chordates, for example, have completely distinct adult body plans, but share many features of early development.

As a result, the Haeckelian vision remains with us, but in a new guise (Nelson 1978, Fink 1982, Patterson 1982). The modern recasting holds that von Baer's law can be interpreted to mean that phylogenetically older (more general) developmental events occur earlier in development, and that newly acquired (more specific) developmental steps generally occur later in ontogeny. This view implies that control events in development operate in a hierarchical fashion, with later events dependent directly on earlier ones. Thus, early development has been commonly regarded as conservative because any change in early development might be expected to affect most subsequent processes (Gould 1977, Raff and Kaufman 1983, Buss 1987, Arthur 1988). A typical statement of this hypothesis is worth quoting. "In complex organisms, early stages of ontogeny are remarkably refractory to evolutionary change, presumably because the differentiation of organ systems and their integration into a functioning body is such a delicate process, so easily derailed by early errors with accumulating effects" (Gould and Lewontin 1979). As will be shown later, it makes a great deal of difference just what "early" means.

It is becoming clear that at least some genetic controls in development operate in a hierarchical fashion; the most convincing cases are the cascade of steps involved in segmentation and determination of segment identity in *Drosophila* (Akam 1987, Ingham 1988) and those controlling sex determination (Hodgkin 1987, 1989). As will be seen below, there is an unwritten corollary to this which states that the heirarchies of early events are global in scope in the developing embryo.

DIRECT DEVELOPING EMBRYOS AS AN EXPERIMENTAL SYSTEM

To go beyond the theoretical considerations discussed above, we need experimental systems in which evolutionary changes in processes of early development can be studied directly and experimentally. The evolutionary origin of direct development within groups characterized by indirect development via a specialized feeding larva offers this possibility. Direct development occurs in such diverse groups as coelenterates, ascidians, mollusks, echinoderms, and frogs. The advantages offered by direct developing systems are (1) that the species generally differ in early development when cell types are few and anatomy simple, (2) that the polarity of evolutionary change is readily determined, and (3) that the organisms to be compared are closely related so that homologies can be established between cell lineages. We have discussed the problems of establishing cellular homologies elsewhere (Raff 1987, Raff 1988, Raff et al. 1989, Wray and McClay 1988, Wray and Raff 1989).

Most species of sea urchins produce, through a sequence of stereotypic developmental events, a characteristic feeding larva called the pluteus. The prospective adult arises later from an anlage called the juvenile rudiment on the left side of the larva (Okazaki 1975). After a few weeks or months of pelagic life, the pluteus settles, and metamorphoses into the juvenile. Some larval cell types are lost at this point (Cameron and Hinegardner 1978).

This pattern of development is primitive for sea urchins (Strathmann 1978), and most species retain it. However, about 20 percent of living species exhibit some degree of direct development (Emlet et al. 1987). Direct development has evolved independently in several orders of sea urchins. Direct developers differ in many details, but share certain features including: large, low density eggs that float in sea water; modifications of cleavage pattern; modifications of blastula formation; incomplete development of the larval gut and absence of larval feeding; and modification or loss of structures typical of the pluteus (Raff 1987). The modified programs of development exhibited by these forms result in accelerated development of adult features.

Our work has utilized two endemic Australian sea urchins which differ in mode of early development. These two species, *Heliocidaris tuberculata* and *H. erythrogramma,* belong to the same genus, and inhabit the same subtidal environment along the shores of New South Wales. However, they differ dramatically in developmental mode. *H. tuberculata* produces a 95 μM diameter egg, which develops via a typical feeding pluteus larva. *H. erythrogramma,* on the other hand, produces a 430 μM diameter egg, which develops into a non-feeding larva that lacks most pluteus features, and undergoes metamorphosis to a juvenile sea urchin in only three and a half days (Williams and Anderson 1975, Raff 1987). Development of a typical sea urchin and of *H. erythrogramma* are compared in Figure 1.

The taxonomic assignment of these species to the same genus, made on the basis of adult morphology, appears to be correct. We have used three molecular indices to estimate phylogenetic relatedness. Comparison of 18S rRNA sequences indicates that these two species differ in only two bases in 1000. This is less than the distance to their nearest examined relative, *Strongylocentrotus purpuratus,* a member of a closely related family (Raff et al. 1988). Similarly, the distance of their single-copy genomic DNA populations T50H is about 15% (Smith et al. 1990). The single-copy DNA distance suggests that divergence occurred 10 to 13 million years ago. Finally, the divergence between mitochondrial DNAs is comparable (McMillan, Palumbi, and Raff, submit-

TYPICAL SEA URCHIN

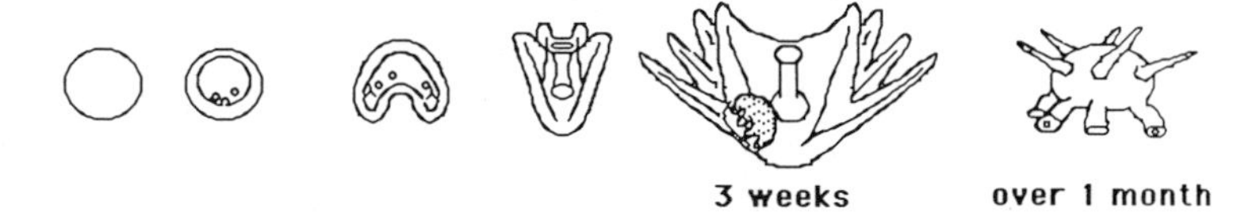

H. ERYTHROGRAMMA

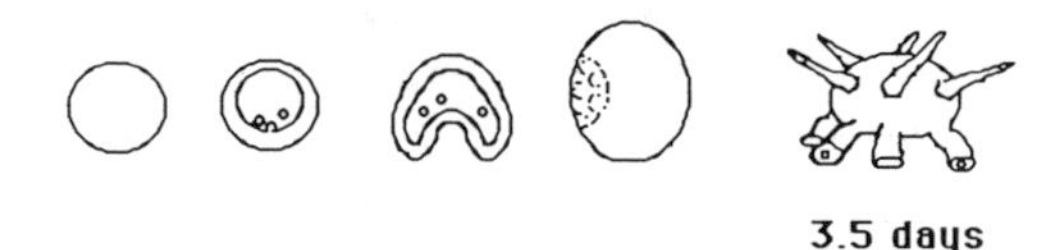

Fig. 1. Patterns of development of a typical sea urchin and the direct developing species *H. erythrogramma.* These drawings are not to scale. The egg of *H. erythrogramma* is four and a half times the diameter of the average typical sea urchin egg. Roughly corresponding stages of development are presented in the two developmental time lines. For the typical developer, these are the egg; mesenchyme blastula with mesenchyme cells entering the blastocoel; gastrula with larval skeleton beginning to be secreted; four armed pluteus larva with differentiated gut and larval skeleton supporting the arms; mature 8-armed pluteus with juvenile rudiment; and newly metamorphosed juvenile. For *H. erythrogramma,* the stages are egg; mesenchyme blastula; gastrula lacking larval skeleton; mature larva lacking arms or functional gut, but with adult juvenile tube feet visible; and newly metamorphosed juvenile. Note the convergence of development in production of the juvenile.

ted). These distances are the same as those measured for congeneric species of the genus *Strongylocentrotus* (Vawter and Brown 1986), confirming their assignment within a single genus. The radical differences in early development arose since these species diverged.

MECHANISMS FOR EVOLUTION OF DIRECT DEVELOPMENT IN A SEA URCHIN

The term *heterochrony* was coined by Haeckel to denote evolutionary changes in the timing of developmental events of a descendent ontogeny relative to the state in an ancestral ontogeny. Heterochrony has been seen as a dominant process of evolutionary change in development (Gould 1977, Raff and Wray 1989), and it offers a useful starting point for interpreting the changes seen between early developmental events in the two *Heliocidaris* species.

Two patterns of change become evident when the relative ordering of events during early development of direct and indirect developing sea urchins are compared (Raff 1987). First, there is an apparent condensation of embryonic development. Thus, *H. erythrogramma* embryos fail to develop some larval structures such as a larval skeleton or gut. Second, the appearance of adult features is initiated early in development, as is their differentiation. These heterochronic changes must be based on underlying changes in patterns of gene expression, cell lineage behavior, and differentiation during early development. A simple heterochrony-based model for the evolution of direct development would predict global changes in those cell lineages that comprise the embryo, such that typical larval differentiation pathways and fates would be condensed or lost, and adult differentiation pathways initiated early in these same cells.

When we began to compare cell differentiation in *H. erythrogramma* with that of indirectly developing species, we found that some cell lineages did indeed behave in this manner. This is the case in skeletogenic mesenchyme cells, which have completely eliminated expression of a larval skeleton, but exhibit an early onset of adult skeletogenesis. This morphogenetic heterochrony is paralleled by a molecular heterochrony in the activity of a gene, msp130, which is expressed only in this cell lineage in both indirect and direct developing embryos (Parks et al. 1988, Wray and McClay 1989). However, we also noted that a striking evolutionary dissociation in evolution among the cell lineages of the embryo had occurred, such that some cell lineages differentiate in the same way as during indirect development, but that differentiation of others, such as the skeletogenic mesenchyme, are greatly modified.

The idea of dissociability of developmental processes derives from Needham (1933), who pointed out that developmental processes are not necessarily causally linked to each other. Thus, during evolution, some developmental processes can be shifted relative to others and still produce a viable, albeit modified, phenotype. Heterochrony is the best known dissociation, but there are certainly others. One result of the dissociability of distinct developmental processes is that organisms often evolve in a mosaic manner. The dissociation of evolution of cell lineages is of particular relevance to arguments on the application of the concept of constraint to early development, and will be discussed later.

CHANGE AND CONSTANCY IN CELL CLEAVAGE PATTERNS AND CELL LINEAGES

Cell lineage plays a central role in fundamental aspects of early development, including establishment of axes and cell type determination. If early development is indeed highly constrained, evolutionary alterations in cell lineage should be rare events during the evolution of a group. This idea finds some support in the early development of spiralian embryos (Raff and Kaufman 1983). Spiralian groups such as annelids, clams and snails, and a few other groups exhibit quite distinct adult body plans, but very similar patterns of cleavage and especially important, conserved cell fates (Wilson 1898, Anderson 1973). In a classic study, Lillie (1898) showed that the evolution of the peculiar larva of the clam *Unio* has involved modifications of early embryonic development. These entailed changes in both patterns and relative rates of cell cleavage such that some embryonic cells that give rise to certain larval structures such as the shell gland are much larger than in the ancestral pattern. However, the overall pattern of spiralian cleavage and cell fate is conserved. This is a pattern of change that at least appears to lie within constrained limits. Whether or not this developmental system is bounded by close constraints has not been demonstrated.

Cleavage patterns and cell fates are conserved among indirect developing sea urchins of all orders (Horstadius 1973, Cameron et al. 1987, Wray and McClay 1988), and one might conclude that cell lineage is a highly constrained aspect of early development among sea urchins. An examination of cell lineages in the direct developing sea urchin *H. erythrogramma* refutes this conclusion. Several very basic aspects of cell lineage have changed in this species with respect to the ancestral condition seen in indirect development. Cleavage is modified in two respects. First, the characteristic unequal fourth cleavage of indirect development is lost, with the fourth cleavage of both animal and vegetal tiers of cells being equal (Williams and Anderson 1975). This change bears upon both the sites of origin of subsequent cell lineages and on the timing of cell lineage fate bifurcations (Wray and Raff 1989). The second change is in cleavage dynamics; synchronous cell cleavages are maintained longer and many more cells are produced (Parks et al. 1988).

Evolutionary modifications in the cell lineage of *H. erythrogramma* are not limited to cleavage patterns. Individual cell lineages have diverged from the ancestral pattern found in typical development in a variety of other ways (Table I) (Wray and Raff 1989). Primary mesenchyme, which forms the calcareous skeleton, shows loss of larval expression, while formation of the adult skeleton is accelerated. In contrast, endoderm omits production of a larval gut without acceleration of adult gut formation. The larval coelom forms, but is much larger and omits an imaginal phase characteristic of typical develop-

TABLE I. Mosaic Evolution of Cell Lineages in Direct Development

Lineage	Typical development	*H. eyrthrogramma*
1. Primary mesenchyme	Larval skeleton	Larval fate deleted, adult accelerated
2. Coelomic precursors	Imaginal, coelom in pluteus	Accelerated, coelom late gastrula
3. Secondary mesenchyme	Larval pigment cells	Larval pigment cells
4. Endoderm	Larval gut	Rudimentary
5. Ectoderm	Larval ectoderm, nervous system, and ciliated band	Larval ectoderm, nervous system, and ciliated band

ment. Other cell types, such as ectoderm, are unmodified, and form the same features as in the pluteus larva. The *H. erythrogramma* larva is thus not merely a deficient pluteus with accelerated adult features. Instead, numerous dissociations have resulted in an entirely novel kind of larva.

Some of the heterochronic modifications in cell lineage pattern in *H. erythrogramma* are accelerations, while others are delays. Thus there is a general delay in the segregation of cell types during early development (Fig. 2). In typical developers, four cells in the 16-cell embryo (micromeres) give rise to two particular cell types (coelomic cells and spicule-forming cells); in *H. erythrogramma,* segregation of these cell types does not occur until at least the 64-cell stage (Wray and Raff 1989; 1990). Concomitant with this delay in lineage segregation, there is an acceleration in the establishment of dorso-ventral asymmetry in *H. erythrogramma.* In species with typical indirect development, the dorso-ventral axis is established at about the 32-cell stage (Kominami 1988), whereas differences in cell fate indicate that this axis is established as early as the 2-cell stage of the direct developer (Wray and Raff 1989, Henry and Raff, 1990).

MOLECULAR HETEROCHRONIES

The strongest evidence for evolutionary conservatism in early development derives from comparative morphology. Many of the processes occurring during early development, however, such as determination and induction, are not manifest morphologically. If early development is indeed evolutionarily conservative, examination of these processes should also show few changes across taxa. In a test of this prediction, gene expression patterns were compared during early development of seven indirect developing sea urchins which exhibit few morphological differences during early development (Wray and McClay 1989). Changes in the spatial and temporal distribution of each of the three tested gene products showed that several molecular heterochronies and heterotopies have accompanied the radiation of sea urchins. At least one additional molecular heterochrony separates the two sub-classes of echinoids (Raff et al. 1984): The primitive cidaroid echinoids express a set of histone genes only in the zygote, whereas euechinoids express this same set of genes both in oogenesis and in the zygote (Angerer et al. 1984). No morphological consequences of changes in timing of expression of this set of genes can be demonstrated (Wells et al. 1986). The existence of non-morphological differences in timing of gene expression among morphologically similar early embryos suggests that morphological evidence for the conservative nature of early development must be interpreted with caution.

Not surprisingly, gene expression heterochronies accompany the evolution of direct development as well. The best documented thus far is that of a gene, msp130, expressed only in skeletogenic mesenchyme in both indirect and direct developing species of *Heliocidaris* (Parks et al. 1988). The msp130 gene exhibits two patterns of expression during indirect development, first in associ-

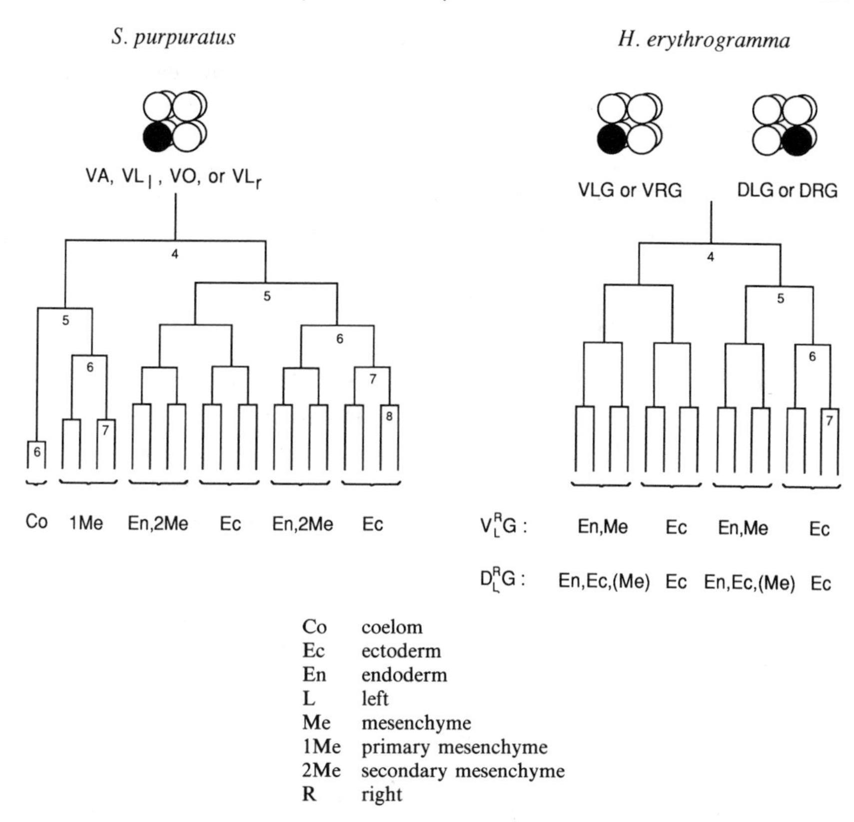

Fig. 2. Comparison of cell lineages in typical and direct developing sea urchins. Lineages of a single vegetal cell at the 8-cell stage are shown for a typical developing species *S. purpuratus* and a direct developing species *H. erythrogramma;* larval cell fates are listed across the bottom. The early asynchrony in division rates of typical development are lost in direct development, as is the asymmetric fourth cleavage. Another asymmetry in cell fates arises even earlier in the direct developer, however: Dorsal and ventral vegetal cells (VLG or VRG vs DLG or VRG) have distinct fates, whereas all vegetal cells (VA, VL_l, VO and VL_r) of the typical developer have equivalent fates. *H. erythrogramma* produces coelom and both primary and secondary mesenchyme, but because the precise origins of these cells are not yet known, Me is used to indicate all mesodermal fates. Fates in parentheses are quantitatively minor and absent from some embryos. Summarized from Cameron et al. 1987, and Wray and Raff 1989.

ation with the larval skeleton, and then in association with production of the adult skeleton. In the direct developing embryo, however, it exhibits only the phase of activity associated with secretion of the adult skeleton; production of the larval skeleton is deleted, and production of the adult skeleton with its corresponding msp130 activity is initiated earlier in development and accelerated.

Changes have also occurred in expression of genes potentially involved in cell determination controls. The uEGF gene encodes a molecule containing tandem epidermal growth factor repeats combined with domains homologous to complement C1s and to avidin (Hursh et al. 1987, Delgadillo-Reynoso et al. 1989). Alternately spliced transcripts containing 21 or 13 EGF domains are present. This molecule has a signal peptide and a pattern of expression suggesting a function in early development in the sea urchin embryo. The gene is evolutionarily conserved, and is expressed in *H. erythrogramma* embryos, but with a different pattern of transcripts from that seen in indirect development (Delgadillo-Reynoso and Raff, unpublished).

REGULATIVE DEVELOPMENT AND EXPERIMENTALLY GENERATED CELL LINEAGES

The classic work of Horstadius (1973) demonstrated that the animal-vegetal axis of the egg is specified prior to fertilization. Further, the three tiers of cells that lie along that axis in the 16-cell embryo have distinct and predictable fates. More recent studies involving the tracing of individual cell fates by injections of fluorescent dyes (Cameron et al. 1987) show that the oral-aboral axis is set up at a constant orientation relative to the first cleavage plane, which always lies along the animal-vegetal axis. These experiments, in combination with studies of cell lineage-restricted gene expression, also demonstrate that cell lineages and their developmental fates, including gene expression patterns, are invariant (reviewed by Davidson 1989).

The paradox is that these embryos, despite their invariant cleavages and developmental patterns, are also highly regulative when experimentally manipulated. Thus, if the first two cells of a sea urchin embryo are separated and cultured, each is capable of producing a half size but morphologically normal pluteus. Even more surprisingly, if a pair of the most animal pole cells (mesomeres, see Fig. 3) of a 16-cell embryo are isolated, in a large number of cases they can produce small larvae with most of the features of plutei (Henry et al. 1989). These cells in the normal course of development produce only ectodermal cells and neurons. Clearly, these regulative properties are of importance to understanding the capacity of early development to evolve.

This point is illustrated by the development of the skeletogenic cells that secrete the larval skeleton. This particular cell type can be produced by three entirely different developmental pathways, two of them experimentally induced (Fig. 3). The first four rounds of cell division produce a 16-cell embryo comprising three distinct tiers of cells (Fig. 3). During normal development, only the micromeres give rise to skeletogenic cells (Horstadius 1973).

If the micromeres or skeletogenic cells are experimentally removed, the macromeres can give rise to skeleton-forming mesenchyme (Fukushi 1962; Horstadius 1973; Ettensohn and McClay 1988). Isolated pairs of mesomeres

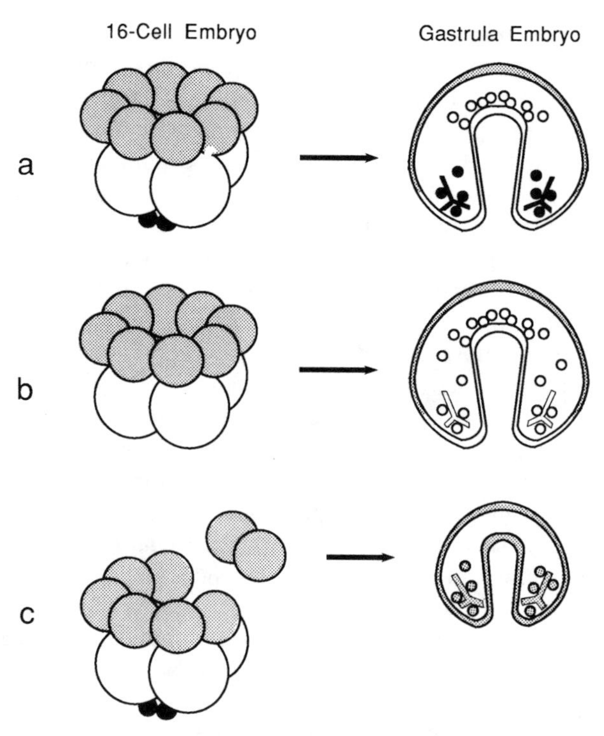

Fig. 3. Three pathways for the formation of skeletogenic mesenchyme cells in sea urchin embryos. **a:** Normal development in which micromeres give rise to a lineage of cells, the primary mesenchyme that exhibit a complex set of behaviors and ultimately secrete the larval skeleton. **b:** Development of embryos from which the micromeres have been deleted experimentally, and cells deriving from the macromeres give rise to skeleton-forming cells. **c:** Development of isolated pairs of mesomeres.

can also generate skeleton-forming mesenchyme cells (Henry et al. 1989) as can mesomeres treated with Li^+ ions (Livingston and Wilt 1989). In the two experimental cases, skeletogenic cells have been produced by novel pathways of cell differentiation, and novel cell lineages have been created. As far as we know these novel pathways have not been exploited as evolutionary modifications of sea urchin development. They do however, indicate that development is very flexible when perturbed, and that evolutionary modifications of cell lineage behavior are possible.

DEVELOPMENTAL STAGE: INTEGRATION AND CONSTRAINT

The extensive modifications in early developmental processes that we have documented in sea urchins clearly establish that early development can evolve in radical ways; they also suggest that it can do so within a relatively short evolutionary time frame. We can now return to the concept of developmental constraints and more accurately assess the evolutionary relevance of von Baer's laws. There are three formal possibilities why any particular evolutionary change in a process during early development might be constrained. They are the following: (1) the novel process cannot be attained by mutation within the existing genetic organization; (2) the novel process is lethal; and (3) the novel process disrupts subsequent developmental processes. All three are valid aspects of the concept of developmental constraint. The second point is of less interest to us than the first and third, because it is essentially a sink for a variety of lethal mutations. Analysis of such mutations can tell a great deal about genetic regulation of specific developmental processes, but probably little about constraints per se.

The ability to attain a novel genetic state might be constrained for quite distinct, but complementary, reasons. According to one view, argued by Arthur (1988) and others, the magnitude of effects of mutations in genes acting early in development are always large because all of development is heirarchically organized. This implies that it would be hard to modify any early developmental processes by small genetic changes. Large evolutionary changes may occasionally be selectively advantageous if they produce fortuitous jumps to a new morphological window, i.e., a macroevolutionary change. According to the second view, an organism has a long evolutionary history in which processes become more interconnected and complex with time, and thus increasingly difficult to modify without disruption (Levinton 1988). One postulated component of this difficulty is that once a gene is introduced into a developmental pathway, it acquires a high degree of interactiveness with other genes and thereafter cannot be easily removed or modified.

The possibility of developmental constraints causing stasis or directional long-term evolutionary trends has been argued from the perspective of development as a genetic program, as well as from the alternate view of epigenetic controls on gene expression and pattern formation (e.g., Arthur 1988 vs. Løvtrup 1974; Levinton 1988). Wimsatt's (1986) concept of "generative entrenchment" and Riedl's (1978) concept of "burden" are similar in supposing that the more heavily subsequent events depend on earlier processes, the more constrained the earlier event is. This idea is an old one and makes good logical sense. The earliest statement of the idea was by Kleinenberg (1886), who suggested that apparently relict structures of early development, such as the notochord of vertebrates, are retained and recapitulated because they participate in the formation of other organ systems in the embryo, and are thus part of the constructional rules of the system. Similar ideas have been presented by Jacob (1977), who pointed out that evolution of new structures is by a process analogous to tinkering; old developmental events or structures are modified for new uses retaining the old processes because they are already there.

The empirical evidence does not bear out a substantial level of developmental constraint in early development (Sedgwick 1894, Garstang 1928, Roth

1984, Elinson 1987, Levinton 1988, Raff 1987, Raff et al. 1989, Strathmann 1988, Wray and McClay 1989). Haeckel's famous drawings of the course of development in vertebrates from fish to human shows that all pass through a morphologically similar pharyngula stage and diverge subsequently in development. However, during development prior to the pharyngula, the differences are profound (Elinson 1987). Eggs, cleavage, gastrulation and germ layer formation are very different in amphibians, birds and mammals, despite their morphologically conserved pharyngulas. Early development in this case actually appears to be less constrained than the later pharyngula stage. The same situation applies among sea urchins, where early development can vary, but all species, even the most extreme direct developers, produce the juvenile rudiment in the same way.

These observations tell us something significant about development and its evolution. Von Baer's laws apply not to the period from the time of fertilization through the establishment of tissues, but to the period from morphogenesis in which the bauplan of the group is visible through late developmental stages leading immediately to the adult. When viewed within this time frame, von Baer's laws are generally valid. What is true is that following the establishment of the bauplan, such as seen in the vertebrate pharyngula or the echinoid juvenile rudiment, most evolutionary changes result in a diverging pattern of later development. In discussions of von Baer's laws with regard to developmental constraints, one has to be careful about what is meant by "earlier."

A reasonable evolutionary description of early development is that, although selective pressures often cause processes and morphology to vary widely within a group, ontogeny converges as the tissues and basic morphology of the bauplan is being laid out (Giard 1904, deBeer 1958, Cohen and Massey 1983, Elinson 1987, Strathmann 1988). It is after this intermediate common condition that ontogeny diverges in late development in accord with von Baer's observations. In terms of potential evolutionary constraints, this says that such constraints are low early, but pronounced later.

We may now be in a position to suggest a cogent mechanistic reason why the evolution of ontogeny is characterized by shifts in developmental constraints with stages in the pattern noted above. Stage A in Figure 4 represents an early stage of development in which constraints are low. Several mechanisms have been suggested (Sander 1982, Strathmann 1988, Raff et al. 1989). Sander (1982) has suggested that redundancies might exist in early developmental processes such that even if a pathway is mutationally deleted, the appropriate state can be attained by an alternate pathway. Such a hypothesis is consistent with the behaviors of complex genetic networks proposed by Kauffman (1987, personal communication 1989), in which a network of interactions can lead to the same final state by more than one pathway. A similar escape from potential constraints arises from the fact that many larval structures are terminal in fate, and do not directly contribute to later development.

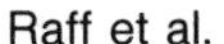

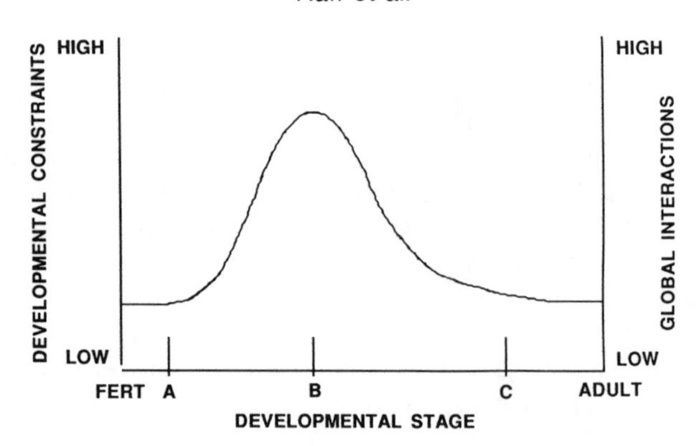

Fig. 4. Variation in level of global interactions and developmental constraints during ontogeny. **Stage A:** Early development, from fertilization to beginning of tissue formation, when few global interactions occur. **Stage B:** An intermediate time in development when global inductive interactions between tissues are at a maximum. **Stage C:** Late development, when global interactions have fallen but local interactions are high.

Thus the larval developmental programs constitute parallel rather than sequential pathways, and those processes involved in producing these features may be dispensible for adult development (Arthur 1988, Strathmann 1988). However, it is clear that this explanation does not explain more than a fraction of evolution of early development because most of early development does directly contribute to laying the groundwork for later ontogeny.

Another property of many early embryos is indicated by empirical observations of the existence of highly regulative development. Regulation indicates a kind of entrainment of development toward establishing axes and cell fates, perhaps analogous to Waddington's (1966) idea of canalization. We are able to visualize regulative abilities of embryos by experimental perturbations (Fig. 3). However, natural variation in cleavage planes and positions of presumptive cytoplasmic determinants also occur and are accommodated in the course of normal development (Henry et al. 1989). Although we know that regulation occurs, we do not yet understand how it works.

We suggest that von Baer's laws and their evolutionary manifestation in developmental constraint might be explained by considering the logic of developmental systems in another way. The three developmental stages indicated in Figure 4 mark early development with low constraints, an intermediate period with high constraints, and finally an interval in late development in which constraints are low again. We propose that constraints are to a large extent a function of the degree of global interaction taking place between components of a developing system. Thus, in early development, constraints exist, but they are relatively low because many early developmental events occur in an epigenetic environment with few interactions, in which cells carry out their differentiative processes with a considerable degree of autonomy. This is not to suggest that interactions do not occur, but rather that they are fewer and less constraining than later. What is critical is that in early development there are redundancies, dispensible pathways, and a lower overall degree of integration between cells and processes.

As the adult body plan is established, interactions increase dramatically. The intermediate stage indicated as B in Figure 4 might represent neurulation in the amphibian embryo. During this stage global inductive interactions are complex and probably at a maximum (Mangold 1961, Slack 1983). The result is the laying out of the tissues and morphology of the pharyngula.

Later stages (C in Fig. 4) might seem to present a paradox in this scheme. Morphology is more complex than in the intermediate stage, yet we are proposing that global interactions are again lower. The basis for this hypothesis is that the period of global interactions takes place when the embryo is still small and undergoing numerous movements of cells in relation to each other. During this period, tissues that are far apart in the adult actually make transient contacts and interact inductively. These conversations are complex, fleeting, and specific. Early induction of lens, for example, requires first the induction of forebrain and eye by preachordal mesoderm. This is followed by sequential inductions involving endoderm, heart mesoderm, and finally retina (Jacobsen 1966). The inductions preceeding that by retina apparently play the major quantitative role, yet these tissues will ultimately lie far from the eye (Henry and Grainger 1987, Grainger et al. 1988).

In contrast, during later development individual structures may be internally tightly integrated, but interact little with other structures. Thus, later eye development has little direct affect on development of limbs or visceral organs. The concept expressed by Riedl as "burden" is well supported by our model for the dependence of developmental constraints on overall integration. Global complexity and burden are at their greatest not in early development, but later.

The study of the relationship between evolution and development provides a unique ability not only to define the evolution of specific developmental processes and their bearing on evolution, but also to use the perspective provided by evolution to infer something of the logic of development itself.

ACKNOWLEDGMENTS

This work was supported by NIH grants RO1 HD 21337 and R01 HD21986 to RAR, and by a NIH postdoctoral fellowship to GAW.

REFERENCES

Akam M (1987): The molecular basis for metameric pattern in the *Drosophila* embryo. Development 101:1.

Alberch P (1982): Developmental constraints in evolutionary processes. In Bonner JT (ed): "Evolution and Development." Berlin: Springer-Verlag, p. 313.

Anderson DT (1973): "Embryology and Phylogeny in Annelids and Arthropods." Oxford: Pergamon Press.

Angerer LM, DeLeon DV, Angerer RC, Showman RM, Wells DE, Raff RA (1984): Delayed accumulation of maternal histone mRNA during sea urchin oogenesis. Dev Biol 101: 477.

Arthur W (1988): "A Theory of Development." Chichester: John Wiley and Sons.

Buss LW (1987): "The Evolution of Individuality." Princeton: Princeton University Press, p. 33.

Cameron RA, Hinegardner RT (1978): Early events in sea urchin metamorphosis: Description and analysis. J Morphol 157: 21.

Cameron RA, Hough-Evans BR, Britten RJ, Davidson EH (1987): Lineage and fate of each blastomere of the eight-cell sea urchin embryo. Genes Dev 1:75.

Cohen J, Massey BD (1983): Larvae and the origins of major phyla. Biol J Lin Soc 19: 321.

Davidson EH (1989): Lineage-specific gene expression and the regulative capacities of the sea urchin embryo: A proposed mechanism. Development 105:421.

deBeer G (1958): "Embryos and Ancestors." Oxford: Oxford University Press.

Delgadillo-Reynoso MG, Rollo DR, Hursh DA, Raff RA (1989): Structural analysis of the uEGF gene in the sea urchin *Strongylocentrotus purpuratus* reveals more similarity to vertebrate than to invertebrate genes with EGF-like repeats. J Mol Evol 29:314.

Eldredge N (1985): "Timeframes." New York: Simon and Schuster.

Elinson RP (1987): "Changes in developmental patterns: Embryos of amphibians with large eggs." In Raff RA, Raff EC (eds): "Development as an Evolutionary Process." New York: Alan R. Liss, Inc., p. 1.

Emlet RB, McEdward LR, Strathmann RR (1987): Echinoderm larval ecology viewed from the egg. In Lawrence J, Jangoux M (eds): "Echinoderm Studies 2." Rotterdam: Balkema, p. 55.

Ettensohn CA, McClay DR (1988): Cell lineage conversion in the sea urchin embryo. Dev Biol 125: 396.

Fink WL (1982): The conceptual relationship between ontogeny and phylogeny. Paleobiol 8: 254.

Fukushi T (1962): The fates of isolated blastoderm cells of sea urchin blastulae and gastrulae inserted into the blastocoel. Bull Mar Biol Stn Asamashi, Tohoku Univ 11: 21.

Garstang W (1922): The theory of recapitulation: A critical restatement of the biogenetic law. J Linn Soc 35: 81.

Garstang W (1928). The origin and evolution of larval forms. Rep British Assoc Adv Sci, Sec D.

Giard A (1904): La poecilogenie. Sixième Congress international de Zoologie, C R Soc Biol (Paris) 6:617.

Gould SJ (1977): "Ontogeny and phylogeny." Cambridge: Harvard Univ Press.

Gould SJ, Lewontin RC (1979): The spandrels of San Marco and the Panglossian paradigm: A critique of the adaptionist programme. Proc R Soc Lond [Biol] 205: 581.

Grainger RM, Henry JJ, Henderson RA (1988): Reinvestigation of the role of the optic vesicle in embryonic lens induction. Development 102: 517.

Henry JJ, Amemiya S, Wray GA, Raff RA (1989): Early inductive interactions are involved in restricting cell fates of mesomeres in sea urchin embryos. Dev Biol.

Henry JJ, Grainger RM (1987): Inductive interactions in the spatial and temporal restriction of lens-forming potential in embryonic ectoderm of *Xenopus laevis.* Dev Biol 124: 200.

Henry JJ, Raff RA (1990): Evolutionary changes in the mechanisms of dorsoventral fate determination in the direct-developing sea urchin *Heliocidaris erythrogramma* Dev Biol 141:55.

Hodgkin J (1987): Sex determination and dosage compensation in *Caenorhabditis elegans.* Annu Rev Genet 21: 133.
Hodgkin J (1989): *Drosophila* sex determination: A cascade of regulated splicing. Cell 56: 905.
Holder N (1983): Developmental constraints and the evolution of vertebrate digit patterns. J Theor Biol 104: 451.
Horstadius S (1973): "Experimental Embryology of Echinoderms." Oxford: Clarendon Press.
Hursh DA, Andrews ME, Raff RA (1987): A sea urchin gene encodes a polypeptide homologous to epidermal growth factor. Science 237: 1487.
Ingham PW (1988). The molecular genetics of embryonic pattern formation in *Drosophila.* Nature 335: 25.
Jablonski D (1986): Larval ecology and macroevolution in marine invertebrates. Bull Marine Sci 39: 565.
Jacob F (1977): Evolution and tinkering. Science 196:1161.
Jacobsen AG (1966). Inductive processes in embryonic development. Science 152: 25.
Kauffman SA (1987): Developmental logic and its evolution. Bioessays 6: 82.
Kleinenberg N (1886): Die Entstehung des Annelids aus der Larve von Lopadorhynchus. Nebst Bemarkungen über die Entwicklung anderer Polychäten. Ztschr Wiss Zool 44: 1.
Kominami T (1988): Determination of dorso-ventral axis in early embryos of the sea urchin, *Hemicentrotus pulcherrimus.* Dev Biol 127: 187.
Levinton J (1988): "Genetics, Paleontology, and Macroevolution." Cambridge: Cambridge Univ Press, p. 217.
Lillie FR (1898): Adaptation in cleavage. In "Biological Lectures of the Marine Biological Laboratory of Woods Hole, Mass." Boston: Ginn and Company, p. 43.
Livingston BT, Wilt FH (1989): Lithium affects the determination of blastomeres in sea urchin embryos. Proc Nat Acad Sci USA 86:3669.
Løvtrup S (1974): "Epigenetics." London: John Wiley and Sons.
Mangold O (1961): Grundzüge der Entwicklungsphysiologie der Wirbeltiere mit besonderer Berücksichtigung der Missbildungen auf grund experimenteller Arbeiten an Urodelen. Acta Genet Med Gemellol (Roma) 10: 1.
Maynard-Smith J, Burian R, Kauffman S, Alberch P, Campbell J, Goodwin B, Lande R, Raup D, Wolpert L (1985): Developmental constraints and evolution. Q Rev Biol 60:-265.
Mileikovski SA (1971): Types of larval development in marine bottom invertebrates, their distribution and ecological significance: A reevaluation. Marine Biol 10: 193.
Needham J (1933): On the dissociability of the fundamental process in ontogenesis. Biol Rev Camb Philos Soc 8: 180.
Nelson G (1978): Ontogeny, phylogeny, paleontology, and the biogenetic law. Syst Zool 27: 324.
Okazaki K (1975): Normal development to metamorphosis. In Czihak G (ed): "The Sea Urchin Embryo," Berlin: Springer-Verlag, p.177.
Oster GF, Shubin N, Murray, JD, Alberch P (1988): Evolution and morphogenetic rules: The shape of the vertebrate limb in ontogeny and phylogeny. Evol 42: 862.
Parks AL, Parr BA, Chin JE, Leaf DS, Raff RA (1988): Molecular analysis of heterochronic changes in the evolution of direct developing sea urchins. J Evol Biol 1: 27.
Patterson C (1982): Morphological characters and homology. In Joysey A, Friday AB (eds): "Problems of Phylogenetic Reconstruction." London: Academic Press, p. 21.
Raff RA (1987): Constraint, flexibility, and phylogenetic history in the evolution of direct development in sea urchins. Dev Biol 119: 6.

Raff RA (1988): Direct developing sea urchins: A system for the study of developmental processes in evolution. In Burke RD, Mladenov PV, Lambert P, Parsley RL (eds): "Echinoderm Biology." Rotterdam: Balkema, p.63.

Raff RA, Anstrom JA, Huffman CJ, Leaf DS, Loo JH, Showman RM, Wells DE (1984): Origin of a gene regulatory mechanism in the evolution of echinoderms. Nature 310: 312.

Raff RA, Field KG, Ghiselin MT, Lane DJ, Olsen GJ, Parks AL, Parr BA, Pace NR, Raff EC (1988): Molecular analysis of distant phylogenetic relationships in echinodermata. In Paul CRC, Smith AB (eds): "Echinoderm Phylogeny and Evolutionary Biology." Oxford: Oxford University Press, p. 29.

Raff RA, Parr BA, Parks AL, Wray GA (1989): Heterochrony and other mechanisms of radical evolutionary change. In Nitecki MH, Nitecki DV (eds): "Evolutionary Innovations." Chicago: University of Chicago Press.

Raff RA, Kaufman TC (1983): "Embryos, Genes, and Evolution." New York: MacMillan.

Raff RA, Wray GA (1989): Heterochrony: Developmental mechanisms and evolutionary results. J Evol Biol, 2:409.

Raup DM (1966): Geometric analysis of shell coiling: General problems. J Paleont 41: 43.

Riedl R (1978): "Order in Living Organisms" (Jefferies RPS, trans). New York: John Wiley and Sons.

Roth VL (1984): On homology. Biol J Linn Soc 22: 13.

Sander K (1982): The evolution of patterning mechanisms: gleanings from insect embryogenesis and spermatogenesis. In Goodwin BC, Holder N, Wylie CC (eds): "Development and Evolution." Cambridge: Cambridge Univ Press, p. 137.

Sedgwick A (1894): On the law of development commonly known as von Baer's law; and on the significance of ancestral rudiments in embryonic development. Q J Micros Sci 36:35.

Slack JMW (1983): "From Egg to Embryo: Determinative Events in Early Development." Cambridge: Cambridge Univ Press, p. 64.

Smith MV, Boom JDG, Raff RA (1990): Single copy DNA distance between two congeneric sea urchin species exhibiting radically different modes of development. Mol Biol Evol 7:315.

Strathmann RR (1978): The evolution and loss of feeding larval stages of marine invertebrates. Evol 32: 894.

Strathmann RR (1985): Feeding and nonfeeding larval life-history evolution in marine invertebrates. Ann Rev Ecol Syst 16: 339.

Strathmann RR (1988): Larvae, phylogeny, and von Baer's Law. In Paul CRC, Smith AB (eds): "Echinoderm Phylogeny and Evolutionary Biology." Oxford: Clarendon Press, p. 53.

Vawter L, Brown WM (1986): Nuclear and mitochondrial DNA comparisons reveal extreme rate variation in the molecular clock. Science 234: 194.

von Baer KE (1828): "Über Entwicklelungsgeschichte der Thiere: Beobachtung und Reflexion." Vol. I. Konigsberg: Borntrager. Translation of Huxley TH, "Fifth Scholium." In Henfrey A and Huxley TH (eds): "Scientific Memoirs, Selected From the Transactions of Foreign Academies of Science, and From Foreign Journals: Natural History." 1853. London: Taylor and Francis, p. 176.

Waddington CH (1966): "Principles of Development and Differentiation." New York: Macmillan.

Wells DE, Anstrom JA, Raff RA, Murray SR, Showman RM (1986): Maternal stores of alpha-subtype histone mRNAs are not required for normal early development of sea urchin embryos. Roux's Arch Dev Biol 195: 252.

Williams DHC, Anderson DT (1975): The reproductive system, embryonic development, larval development and metamorphosis of the sea urchin *Heliocidaris erythrogramma* (Val.) (Echinoidea: Echinometridae). Aust J Zool 23: 371.

Wilson EB (1898): Cell lineage and ancestral reminiscence. In "Biological Lectures: The Marine Biological Laboratory of Woods Hole, Mass." Boston: Ginn and Company, p. 21.

Wimsatt WC (1986): Developmental constraints, generative entrenchment, and the innate acquired distinction. In Bechtel W (ed): "Integrating Scientific Disciplines." Dordrecht: Martines-Nijhoff, p. 185.

Wray GA, McClay DR (1988): The origin of spicule-forming cells in a "primitive" sea urchin *(Eucidaris tribuloides)* which lacks primary mesenchyme cells. Development 103: 305.

Wray GA, McClay DR (1989): Molecular heterochronies and heterotopies in early echinoid development. Evolution 43:403.

Wray GA, Raff RA (1989): Evolutionary modification of cell lineage and fate in the direct developing sea urchin *Heliocidaris erythrogramma.* Dev Biol 132: 458.

Wray GA, Raff RA (1990): Novel origins of lineage founder cells in the direct-developing sea urchin *Heliocidaris erythrogramma*. Dev Biol 141:41.

New Perspectives on Evolution, pages 209-223

Our Recent African Past: New Mitochondrial Perspectives on Human Evolution

REBECCA L. CANN
Department of Genetics, University of Hawaii at Manoa, Honolulu, Hawaii 96822

> At any given moment there is an orthodoxy of ideas, which it is assumed all right-thinking people will accept without question. It is not exactly forbidden to say this, that, or the other, but it's "not done" to say it, just as in mid-Victorian times it was "not done" to mention trousers in the presence of a lady. Anyone who challenges the prevailing orthodoxy finds himself silenced with surprising effectiveness.
>
> **G. Orwell (1972), suppressed preface to *Animal Farm*, published posthumously.**

INTRODUCTION

Creatures of Our Past

We are working today with the biggest conceptual change and scientific challenge in human evolutionary studies since the description of an Australopithecine fossil published by Raymond Dart in 1925. In the next ten years we will celebrate the progress of molecular biologists who are now deciphering the human genome, base-pair by base-pair. The genetic basis of many morphological traits will be established. Technology for recovering DNA from fossils will improve, and clearly defined phylogenetic hypotheses for the evolution of genes in populations will be proposed. Organismal biologists will then be called on to assess the ultimate significance of far more sophisticated information about human evolution than that which exists today. How will they react? If the present is any guide to the future, the interactions between molecular and organismal scientists concerned with human evolution will be bumpy, and the interchange between disciplines best characterized as one of missed opportunities. My perspective on this problem stems from controversy surrounding the evolution of mitochondrial DNA in humans.

Until the mid 1960s, human evolutionary biology was considered the primary domain of paleontologists, anatomists, and archaeologists. It expanded somewhat in the '60s at the urging of Louis Leakey to include and embrace primatology. The distinction is important, for primatologists were invited into the fold by the prevailing experts, who were curious about the close similarity of human and ape behavioral patterns and the significance of this data for inferences made about the evolution of human social behaviors. Together, they constructed their vision of a day in the life of a typical Australopithecine. Drawing on data gathered from chimpanzee, gorilla, and baboon social groups, they generalized a scenario for a series of gradual morphological and behavioral changes our early human ancestors, leading to the evolution of the genus *Homo.* These included the use of a home base, a division of labor between monogamous males and females, and a defined group of foraging strategies linked to the use of stone or bone tools (Isaac and McCown 1976).

Shifting Paradigms

During this era, molecular anthropology also came of age (Goodman and Tashian 1976). The exclusive concern of biochemists and geneticists was antibodies, proteins, and later, the DNA of modern species. At stake was the contention that a detailed understanding of process and pattern in human evolution would primarily come from studies of morphology and behavior as reflected in the fossil record. Some evolutionary biologists felt that the concentration on the genetic material of modern species would only help fill in the blanks defined by other disciplines.

Molecular anthropologists were not from the chosen few and often had a combative personal style, reminiscent of cowboy molecular biologists. As a group, they asserted the intellectual superiority of their approach, founded in the direct comparison of genes, over anything used in the past to study evolution. They failed, however, to unite behind a single interpretation of their results. Some were armed with data that did not fit the time scales suggested by the fossil record without special adjustments made for unequal rates of evolution. Although it was possible to imagine that natural selection could result in different rates of change in disparate groups, the demonstration that rates varied was ad hoc. Instead, some disputed the assignment of particular Ramapithecine or Sivapithecine fossils to the human lineage, and the dating of the very fossils. The genera of Miocene fossil apes in question are now considered to be more closely tied to the evolution of orangutans than humans (discussed at great length in Ciochon and Corruccini 1983; Andrews 1987). However, fifteen long, contentious years passed before the contributions of molecular biology to human evolution were fully recognized.

A molecular perspective in human evolution is necessary because transitions between species become impossible to model from fossils alone without the correct phylogenetic tree (White et al. 1981), a point that is essential to the understanding of mtDNA data. Because the time of divergence between humans and African apes is so short, 5 plus or minus 2 million years (Brown et al. 1982; Sibley and Ahlquist 1984), Australopithecines dating from approximately 4 to 1.5 million years before present (bp) are key fossils. They are not just some smaller-brained version of the average suburbanite American, operating in the African landscape without a Sharper Image catalog. Their unique place in human evolution as representatives of some basal unit was never in question, but the full zoological perspective of ourselves as an evolutionary lineage derived from a limited subset of them was. All species identified in this early adaptive radiation would face extinction, and only one would be ancestral to a modern taxonomic unit (Gould 1987).

One additional point can be abstracted. With species capable of observational learning, cultural innovations can drive morphological change at unpredictable rates (Wyles et al. 1983). Anthropologists had to reckon with the knowledge that our bodies were constrained by this short geological time scale for evolutionary change, less than one-third that originally proposed by paleontologists. To paraphrase Pascal, both the beast and the greatness were seen to be united in the Australopithecine fossils. Some modern human behaviors and associated morphologies were likely to be ancient and shared with African apes, because the DNA sequences of genes controlling their expression were so similar. The search for morphological evidence of behaviors that were truely unique to the human lineage, such those connected with speech and bipedalism, took on new significance (Holloway 1983). Molecular dates de-modernized Australopithecines and allowed us instead to focus instead on the ape-like qualities that they truly expressed. I wondered as a beginning graduate student what would happen when the molecular methodologies were applied instead to the more recent side of human evolution, and directed at uncovering the roots of modern people.

NEW CONTROVERSIES STEMMING FROM MTDNA

Some evolutionary biologists now believe that once again, the time scale for human evolution needs readjusting. At the same time, we are also reconsidering hypotheses about the geographic origin of our own species. Many of us interested in the later stages of human evolution are struggling to synthesize new information about the antiquity of genetic lineages, language roots, and fossils assigned as anatomically modern people in certain geographic regions of the Old World. We contend that human maternal genetic lineages discovered using mtDNA from modern populations predate the appearance of modern phenotypes preserved as fossils in Africa and the Middle East approximately 100,000 years ago.

Anthropologists had previously asserted as late as 1978 that *Homo sapiens sapiens,* the taxonomic designation given to anatomically modern people, evolved in the Middle East 40,000 years ago. Mitochondrial DNA ten years later suggests instead that the origins of modern people may be found in sub-Saharan Africa. I now believe, as do some paleoanthropologists, that this hypothesis best accounts for the earliest occurrence in East Africa and South Africa of anatomically modern people, judged from fossil evidence (Stringer and Andrews 1988). What has been more troublesome is the contention that all lineages of modern people trace their mitochondrial DNA to a common maternal ancestor, the one lucky mother, present in Africa perhaps 200,000 years ago. This part of the argument has now become incredibly confounded with religious fanaticism, racism, bad population biology, and a failure to remember that the evolutionary process necessarily entails the transmission of genes from parent to offspring in an unbroken line.

Exactly how the change from archaic (a classic weasel-word) phenotypes to those seen in people today took place is the question (Howells 1981). Was there a slow, gradual change in all geographic groups dispersed around the world, or was the change sudden, involving the actual replacement of one group of humans by another? Unlike the last big debate generated by molecular data, we have scientific advantages on a number of fronts, and perhaps it will not take scientists fifteen years to fully understand and evaluate the model of replacement now proposed. There are three specific reasons for holding such an optimistic view.

First, we understand the appropriate archaeological context in which many artifacts associated with the fossils were made and used. Living sites may actually be sites of human habitation (Toth and Schick 1986), as opposed to the random accumulation of bones and tools by natural water forces or direct accumulation by animal scavengers that accounts for much of the earlier material (Binford 1981). Tools associated with particular fossils may have actually been made and used by those representative individuals or their close genetic relatives. A second advantage in examining this period of the evolutionary record is that the bulk of fossil material potentially germane to any discussion is increased by many orders of magnitude, the closer we come to the present. Fossils are rare and precious materials, and there is a natural tendency to stress the unappreciated uniqueness of each newly discovered one. With the later stages of human evolution, there are more fossils to consider, as well as innumerable presumed modern descendents of both sexes, all age groups, in all stages of nutrition and health. Thirdly, in the near future, it will be possible to extract DNA sequences directly from old bones, as we now do from mummified materials that are 3000 years old (Pääbo 1989).

OLD PROBLEMS

There is no actual population of fossil humans, in the true sense of the word, available for our analysis that reflects the evolutionary processes operating on a modern gene pool, or deme, until long after the evolution of modern people. The exclusive but necessary emphasis on rare fossil evidence therefore requires a periodic attitude adjustment that is forced on students of human evolution.

It usually comes from zoologists operating under the familiar taxonomic rules used to study modern animal species, and is necessary in order to reassert the role of populational thinking and the neoDarwinian synthesis into human systematics (Mayr 1950). Dispersal, migration, assortative mating, and gene flow have become primary concerns of evolutionary biologists since the 1950s. In order to model these forces, we make assumptions about their operation on paleontological populations and in doing so project our own cultural biases onto technologically simple people, especially in regard to mate choice and preferred population densities. It is impossible to avoid these mistakes. They will only decrease in number when scientists of different gender and ethnic backgrounds become involved in the modeling process, and they give us some starting point from which to modify later hypotheses.

Even with new DNA sequences, new fossils, and new dates for old fossils, we are still puzzled by the latest stages of human evolution. Today we are currently able to resolve just the broadest outline of a dispersal event that traces back to the maternal ancestor of us all (Wilson et al. 1987). I will now turn to the nature of mitochondrial DNA (mtDNA) studies in humans, point to problems in our interpretation or understanding of this evidence, and relay how traditional anthropology might be accommodated in a cohesive picture of recent human history.

THE MITOCHONDRIAL SYSTEM

A Simple Tool for Evolutionary Biologists

In 1987, Mark Stoneking, Allan Wilson, and I published a maximum parsimony tree of modern humans based on character state data from restriction site analysis at about 440 sites in 148 distinct mitochondrial DNA genomes (Cann et al. 1987). The work reported stood on the shoulders of others, and is continuing today in the laboratories of many researchers. It is and was possible because of our knowledge of the complete sequence of one human mitochondrial genome (Anderson et al. 1981).

MtDNA is contained within mitochondria in the cytoplasm, and is a simple, small molecule coding for 37 genes used in the production of cellular energy (Attardi and Schatz 1988). It has two major advantages when it comes to studying evolution. First, the nucleotide sequence of genes in this subset of the genome changes very fast in mammals, about 10 times faster than of a gene of a comparable functional class which would be housed in the nucleus of the cell (Brown et al. 1982). Second, these genes are inherited maternally in animals, and are not subject to the vagaries of sexual recombination at every meiosis. Barring mutation either in somatic tissues of progeny (Wallace et al. 1988) or in the maternal gametes themselves, the offspring of a particular female will inherit her mtDNA sequence. Excellent review articles have appeared in the last few years that discuss data bearing on these points (Harrison

1989, Moritz et al. 1987). In addition, the widespread use of the polymerase chain reaction, or PCR, will allow the specific testing of apparent exceptions to these generalizations (Kocher et al. 1989).

Outlines of the Human Mitochondrial Tree

Tradition in the 1970s stated that anatomically modern people were less than 50,000 years old, and that there was a near simultaneous appearance of modern morphology on all geographic continents, along with the enigmatic survival of ineages labeled Neanderthal in Northern and Central Europe until about 30,000 years bp. What we expected to see was a phylogenetic tree that would reflect this assertion, because protein and associated red-cell polymorphisms assayed by electrophoretic methods were thought to be consistent with the hypothesis (Cavalli-Sforza 1974).

An alternative to this interpretation had been suggested by Masatoshi Nei and colleagues, who used the foundations of the neutral theory to study levels of genetic heterozygosity in human populations. Nei and his group partitioned variance between and within populations to show that Africans could represent the ancestral geographic source from which later groups arose, and the age of human races, grouped as Asians, Africans, and Europeans, would be older than 40,000 years (Nei and Roychoudhury 1982). Limited studies with a small number of mitochondrial restriction sites by Cavalli-Sforza's group were consistent with an African origin, but that alternative was not favored (Johnson et al. 1983).

Our original tree—and ones produced since then by additional computer algorithms with alternative assumptions about rates of fixation along lineages—favors the evolution of an ancestral population in Africa, followed by a migration from Africa, instead of the in situ evolution of modern people from separate geographic origins. A simplified version of this appears in Figure 1. It shows an African base either by midpoint rooting, where the root of a phylogenetic tree is drawn between the two most divergent lineages, or by outgroup rooting, where a species is used that is known to be more distant to the groups of interest than they are to themselves. We take the broad perspective that such an interpretation is consistent with conclusions drawn from globin gene polymophisms and 40 additional nuclear loci (Kidd et al. 1989), but that it is impossible to derive a time scale with sufficient resolution so far using nuclear genes, because of the complications of linkage. Confirmation of the African-based pattern will be seen in Linda Vigilant's work with mtDNA of pygmies and !Kung (Vigilant et al. 1989) and Tom Kocher's (Kocher and Wilson 1989 and in preparation) analysis of the control region (D-loop) sequences in humans and chimpanzees using the Polymerase Chain Reaction (Saiki et al. 1985).

If the pattern shown here is so self evident, why is it that reasonable people dispute this finding? In the early days of restriction mapping, there was an

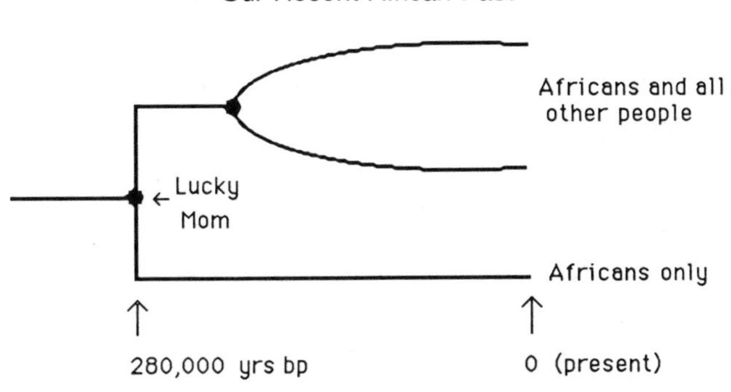

Fig. 1. This diagram portrays a simplified view of the human mitochondrial DNA tree, obtained from both restriction site and direct sequence analysis. The last maternal ancestor is inferred to have lived approximately 280,000 years ago. The tree shows two main branches, corresponding to a split between one clade that contains only people whose maternal ancestors were Africans, and a second clade containing Africans and the rest of the world's people.

extreme bias in sites reported as variable in the human population. Interest in finding differences between populations led to the selected use of restriction enzymes that showed expected results, and there was a general under reporting of the shared polymorphisms that failed to generate these distinctions. In fact, they form the bulk of the polymorphic restriction sites known in humans (Stoneking and Cann 1989). Other authors who dismissed the African origin hypothesis as a myth, selected Africans from the second, more highly derived branch of the original tree, and did not even consider the consequences of drawing an African lineage from the first branch (Excoffier and Langaney 1989). The wisdom of using a predominantly American Black population to represent African genotypes was also questioned (Wainscott 1987). Finally, some scientists objected for purely statistical reasons that the standard error in the distances measured was so large that virtually any pattern would be consistent with the data (Saitou and Omoto 1987). These objections can and have been answered by sequence data from additional samples (Vigilant et al. 1989).

WHEN DID THE "LUCKY MOTHER" LIVE?

Estimates of the time since our species last shared a common, maternal ancestor have varied. We attempted to be conservative in our original suggestion of 140,000–280,000 years. This was to acknowledge that the error associated with correctly placing the root on the restriction site tree could be large, due to our inability to resolve multiple substitutions along certain branches. Internal calibration of the human tree, based on divergence between highland New Guinea and Aboriginal Australian populations, gave some confidence that this date could be quite old (Stoneking et al. 1986). The use of actual sequences instead of restriction sites brings a level of refinement to the data, because many more changes can be counted, and fewer assumptions must be made about the number and type of base substitutions responsible for a restriction site change. Potential confusion with length mutations and confirmational substitutions was not an issue in the first tree, because no length mutations were ever included in the character state data due to uncertainty of their map position and the way they could be generated. Researchers now suggest that the roots of human mtDNA evolution can be estimated at about 200,000 years bp, on the basis of D-loop sequences from 84 individuals (Vigilant et al. 1989). This agrees with the estimate based on our earlier restriction site surveys.

OUR LUCKY MOTHER COULD HAVE HAD AN ARCHAIC PHENOTYPE

We are unable to state on the basis of mitochondrial data alone what the last common maternal ancestor of all modern humans looked like. There are no known anatomically modern fossils in any region of the world as old as 280,000 years bp. Paleoanthropologists have identified a series of transitional forms that date to this time, fossils which have some characteristics of older, *Homo erectus* populations, but at the same time show the expansion of the braincase seen in *Homo sapiens* (Bräuer 1989). The specimens grouped definitively as *Homo sapiens* from this time are present but are not yet anatomically modern, showing dental and cranial features that set them apart from all people alive today. Some researchers suspect that they will ultimately link these archaic fossils to African populations (Stringer 1990). This view is the opposite of that held by the regionalists, who claim that isolated subpopulations of archaic *Homo sapiens* preserved minimal species contact by gene flow and gave rise to their own separate modern groups. In the Far East, *Homo erectus* skeletons persist roughly to 290,000 years bp (Wolpoff 1989), giving some credence to the notion that human evolutionary change was not a geographically uniform process.

THE SPREAD OF MODERN LINEAGES

Based on the average number of differences between every pairwise combination of humans from the 148 mtDNA genome set, we estimated that two people drawn at random would differ at about 9 restriction sites. The distribution was slightly skewed but approximated a normal curve. One inference that can be drawn from the shape of the curve is that the human population from which it was taken is a compound group, containing both older and new lineages.

Lineages with the most changes at first glance are usually African, a consequence of their greater age. (It would also be possible to imagine that African lineages have higher mutation rates, and this possibility was the one that led Johnson et al. (1983) to discount their data for an African origin of all modern people.) According to expectations of random extinction, older populations should have fewer surviving lineages, a phenomenon John Avise and colleagues have called lineage sorting (Avise et al. 1988). As we go back in time, more lineages are lost from the population due to the probability that at some step only sons were produced or that no female offspring survived to reproduce. If modern European or Asian populations were derived from ancestral lineages dating to the isolation of *Homo erectus* populations on these continents, we would expect that their age profiles would at least match that of African populations, and this is not the case (Wilson et al. 1987).

Fewer surviving lineages do not imply that a bottleneck in absolute population size necessarily took place in the past, and it is essential to consider the effects that rapid growth and high fecundity have on the survival of lineages (Cann 1988). Populations of a relatively younger age, stemming from demographically expanding groups, should have a larger number of lineages that are separated by only a few changes. They will differ from the patterns seen in populations that have reached equilibrium, are practicing female infanticide, or are experiencing a demographic catastrophe due to the introduction of disease.

New dates for fossils associated with deposits at Qafzeh and Es Skhul in the Middle East, almost 100,000 years bp, suggest that Africa and southwest Asia acted as a refuge for human hunter-gatherers during the Pleistocene. If we imagine parts of Europe or Asia becoming less attractive due to glaciation, it is entirely possible that some humans retreated to warmer climates following the rivers that crossed the Sahara and are now mapped through satellite photos. Accumulation of stone tools along these waterways and their close proximity to regions known to contain anatomically modern people makes North Africa and the Middle East look suspiciously like a corridor for dispersal to Europe and parts of Asia.

The shape of lineage distribution curves in populations confined to one geographic area are the basis of new claims by Wilson's group (A.C. Wilson, personal communication) that the actual spread of Cro Magnon people as invading Europeans can be detected in the mitochondrial tree. For well-defined populations such as Sardinians and Scandinavians, the proliferation of lineages found in these groups can be traced to a common maternal ancestor approximately 70,000–40,000 years ago. If more bushes within the worldwide tree can be associated with a uniform time range, one of the most persistent questions in human evolution, that of the fate of Neanderthals, may be answered. As the data for Europeans now stand, both the introgression and replacement models are plausible.

THE FATE OF OLDER HUMAN LINEAGES

While our ancestors were spreading from their African homeland, they must have come in contact with at least some of the earlier humans that are known as fossils from China, Europe, North Africa, and Java. Sufficient morphological differences exist between modern people and *Homo erectus* populations that one would suspect assortative mating to operate and keep these groups as separate taxonomic units. Reproductive isolation between subspecies may have been incomplete, however, in the case of Neanderthals and modern people. Archaeological evidence that Neanderthals actually buried their dead, practiced ritual behavior, or engaged in other symbolic communication has now been questioned (Gargett 1989).

Today there is a move among some anthropologists to recognize more formally the cultural gulf which may have existed between archaic and anatomically modern people in Western Europe during the last 100,000 years (Mellars 1989). If Neanderthal fossils are allocated their own species designation, it will be due largely to the recognition that cultural change has so accelerated modern human evolution as to render morphology a useless yardstick for defining the limits of gene flow between populations. New suggestions have been made that modern languages demonstrate the unity of their common origin from an African base (Cavalli-Sforza et al. 1988) and that European languages still contain elements that reflect ancient genetic differences between speakers (Harding and Sokal 1988). The further turf invasion of linguistics by biologists will ruffle some feathers but stimulate a larger community of researchers to look for specific predictions of the temperate zone extinction and African recolonization model for the evolution of *Homo sapiens sapiens.*

NEW DIRECTIONS USING MTDNA AND THE POLYMERASE CHAIN REACTION

Recent Human Expansion to New Habitats

Direct sequencing of PCR-amplified mitochondrial genes from hair samples is now in progress for a number of human populations that only recently colonized certain areas of the world. We are in need of specific information from these populations that are fast disappearing due to diseases of the newly adopted sedentary lifestyle. Other small populations are being lost as distinct ethnic entities through assimilation into the global network of human culture. The efforts of human geneticists such as Ken Kidd at Yale University to establish a bank of permanent cell lines from aboriginal people is one response to the growing need for DNA from a variety of populations. Comparative studies help us form a more complete idea of what sequence diversity in the human genome is all about, and research emphasis on the most recent parts of the mitochondrial tree will help fill in gaps in our knowledge about how fast new lineages are generated and lost in expanding populations.

A new alternative to the sample collection problem is DNA in cells attached to hair follicles or in the hair shaft itself (Higuchi et al. 1988), which can be collected by anyone from remote populations. It obviates the need for liquid nitrogen, needles, or special handling permits. Using specific mtDNA primers for the 12s rRNA subunit gene, a section of the Cytochrome B gene, a small non-coding region designated region 5, and the most variable portions of the D-loop (Fig. 2), it is now possible to obtain enough mtDNA to sequence genes and parts of genes from many donors who would find it culturally distasteful, dangerous, or uncomfortable to give blood. Our study of the mtDNA diversity in Native Hawaiian populations (Cann and Rickards 1989) has so far given us information that the founding of Hawaii by Polynesians 2,000 years ago involved at least four unrelated female ancestors (Fig. 3) in the first canoe, presumably sailing north from the Marquesas (Kirch 1984). In addition, a length mutation in a non-coding region between the tRNA Lysine gene and the Cytochrome Oxidase subunit 2 gene that has been called a marker for Polynesian gene pools is present in about 80 percent of the Hawaiians examined so far. Others are tracing the drift of this marker in related Polynesian societies for evidence of admixture between Polynesians, Melanesians, and Europeans (Herztberg et al. 1989)

Disease, PCR, and Ancient Lineages

One of the most exciting applications of new technology to human evolution will be the actual study of the non-random loss of specific maternal lineages in response to the introduction of novel selective forces. Successive population

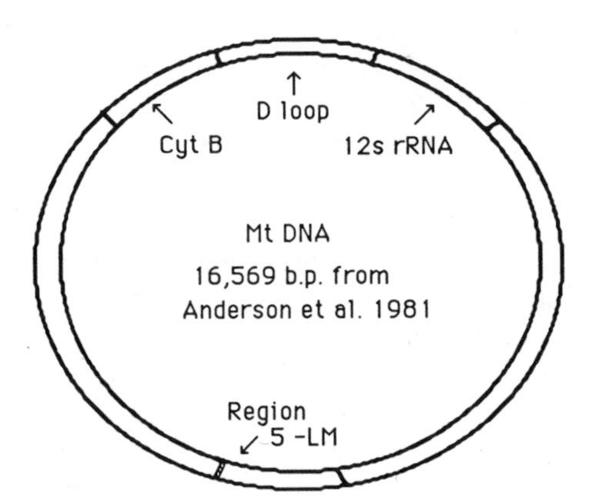

Fig. 2. The mitochondrial genome is shown as a double-stranded circular molecule, with the regions easily amplified using the Polymerase Chain Reaction and universal primers described by Kocher et al. 1989. Cyt B refers to the cytochrome B gene, D-loop is the control region of the molecule containing the heavy strand origin of replication, and 12s rRNA is the small ribosomal subunit of the mitochondria. Region 5 corresponds to a small, non-coding region of approximately 20 base pairs between the tRNA Lysine gene and the Cytochrome Oxidase subunit 2 gene, which shows length polymorphism in humans.

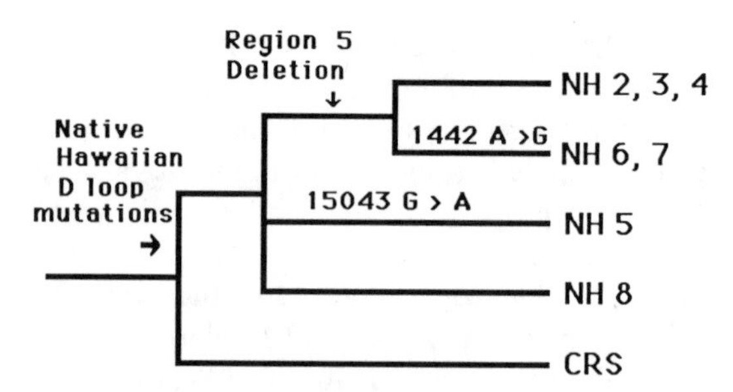

Fig. 3. Using PCR-amplified mtDNA from total genomic DNAs, we have identified at least four separate female lineages among native Hawaiian donors from four islands. D-loop mutations include deletions at 15982 and 16022, along with insertions at 16050 and 16093, relative to the Anderson et al. sequence referred to here as CRS.

bottlenecks reduce the level of genetic variability in an isolate, and a short time scale makes it unlikely that new mutations will have enough time to appear or be fixed in that group. The end result is a genetically homogeneous population with reduced long-term fitness.

Hawaii unfortunately represents an excellent demonstration of this process operating in the last 200 years, stemming from the arrival of European discoverers. The Native Hawaiian population today is one-twentieth its original size before contact with Europeans. Initially, it grew from less than 150 individuals to 400,000 people in about 2,000 years. British and American ship crews remarked originally on the general robust health and appearance of Hawaiians, noted for their stature and girth. This was to change within two generations of contact with the outside world and its diseases. The two population bottlenecks that Native Hawaiians experienced may be currently contributing to a high incidence of health problems with a strong genetic component, such as cancer, diabetes, bronchial asthma, and club foot.

Today, we are engaged in a study utilizing the remains of ancient Hawaiian donors preserved in gentle sand burials, lava tubes, and boggy ground near estuaries. Their mitochondrial genes will be compared to those of modern donors, to document the survival or turnover of lineages in a known period of time. Additionally, we will be searching for the correlations between the introduction of disease pathogens and the survival of specific lineages at the expense of others. The strength of natural selection is notoriously difficult to demonstrate or quantify (Endler 1977), and this approach may be the most straightforward test possible in humans. One can imagine a whole new field of paleoepidemiology arising from PCR applied in this fashion. If so, the issue of reburial of archaeological remains by universities and museums takes on new scientific significance.

CONCLUSIONS

This work would not have developed to its current maturity without the input of interested molecular and organismal biologists. It feeds on the technology developed by molecular researchers, and contributes back to them the knowledge that a variety of sequences are compatible within a complex metabolic pathway that was originally assumed not to vary much, because of functional constraints on essential, individual subunits. It also depends on the judgment of morphologists for placing correct estimates on fossils to calibrate the time scale of mutational change, but gives added depth to their interpretations of population variability. Goodwill and the understanding that both camps are driven by a desire to understand the evolutionary process in as wide a scope as possible will be required for further progress. Without this tolerance, we merely demonstrate that there is luck, as well as some art, in picking those problems that are solvable.

ACKNOWLEDGMENTS

This paper is dedicated to Vincent M. Sarich, teacher and advocate of unpopular crusades. I thank W. Brown, L. Freed, R. Higuchi, T. Kocher, M. Pietrusewsky, O. Rickards, M. Stoneking, L. Vigilant, A. Wilson, and members of my lab for information and discussion. Some DNAs from Native Hawaiians were gathered by Stuart Newfeld. The work reported here was assisted by grants from the National Science Foundation (A. Wilson and R. L. Cann) and a Matsuda Fellowship from the University of Hawaii (R. L. Cann).

REFERENCES

Anderson S, Bankier AT, Barrell BG, DeBruijn MHL, Coulson AR, Drouin J, Eperon IC, Nierlick DP, Roe BA, Sanger F, Schreirer PH, Smith AJH, Staden R, Young IG (1981): Sequence and organization of the human mitochondrial genome. Nature 290: 457–465.

Andrews P (1987): Aspects of hominoid phylogeny. In Patterson C (ed): "Molecules and Morphology in Evolution: Conflict or Compromise?" New York: Cambridge Univ Press, pp 22–53.

Avise JC, Ball RM, Arnold J (1988): Current versus historical population sizes in vertebrates species with high gene flow: A comparison based on mitochondrial DNA lineages and inbreeding theory for neutral mutations. Mol Biol Evol 5:331–344.

Attardi G, Schatz G (1988): Biogenesis of mitochondria. Annu Rev Cell Biol 4:289–333.

Binford L (1981): "Bones: Ancient men and modern myths." New York: Academic Press.

Bräuer G (1989): The evolution of modern humans: A comparison of the African and non-African evidence. In Mellars P, Stringer CB (eds): "The Human Revolution: Behavioural and Biological Perspectives on the Origins of Modern Humans." Princeton, NJ: Princeton Univ Press, pp 123–154.

Brown WM, Prager EM, Wang A, Wilson AC (1982): Mitochondrial DNA sequences of primates: Tempo and mode of evolution. J Mol Evol 18:225–239.

Cann RL (1988): DNA and human origins. Ann Rev Anthropol 17:127–143.

Cann RL, Rickards O (1989): Human mitochondrial DNA mutations in Polynesians detected rapidly by PCR amplification. J Cell Biochem Suppl 13C:128.
Cann RL, Stoneking M, Wilson AC (1987): Mitochondrial DNA and human evolution. Nature 325:31–36.
Cavalli-Sforza LL (1974): The genetics of human populations. Sci Am 231:80–89.
Cavalli-Sforza LL, Piazza A, Menozzi P, Mountain J (1988): Reconstruction of human evolution: Bringing together genetic, archeological, and linguistic data. Proc Natl Acad Sci USA 85:6002–6005.
Ciochon RL, Corruccini RS (eds) (1983): "New Interpretations of Ape and Human Ancestry." New York: Plenum Press.
Endler JA (1977): "Geographic Variation, Speciation, and Clines." Princeton, NJ: Princeton Univ Press.
Excoffier L, Langaney A (1989): Origin and differentiation of human mitochondrial DNA. Am J Hum Genet 44:73–85.
Gargett R H (1989): The evidence for Neanderthal burial. Curr Anthro 30(2):157–189.
Goodman M, Tashian RE (eds) (1976): "Molecular Anthropology." New York: Plenum Press.
Gould SJ (1987): Bushes all the way down. Nat Hist 96(6):12–19.
Harding R, Sokal RR (1988): Classification of the European language families by genetic distance. Proc Natl Acad Sci USA 85:9370–9372.
Harrison RG (1989): Animal mitochondrial DNA as a genetic marker in population and evolutionary biology. Trends Ecol Evol 4(1):6–11.
Hertzberg M, Michelson KNP, Serjeantson SW, Prior JF, Trent RJ (1989): An Asian-specific 9-bp deletion of mitochondrial DNA is frequently found in Polynesians. Am J Hum Genet 44:504–510.
Higuchi R, von Beroldingen CH, Sensabaugh GF, Erlich HA (1988): DNA typing from single hairs. Nature 332:543–546.
Holloway RL (1983): Cerebral brain endocast pattern of *Australopithecus afarensis* hominid. Nature 303:420–422.
Howells WW (1981): Current theories on the origin of *Homo sapiens sapiens.* Colloques internationaux du Centre National de la Recherche Scientifique (France) No. 599, "Less processus de l'hominisation" pp 73–77.
Isaac G, McCown ER (eds) (1976): "Human Origins: Louis Leakey and the East African Evidence." Menlo Park, CA: W. A. Benjamin Inc.
Johnson MJ, Wallace DC, Ferris SD, Rattazzi MC, Cavalli-Sforza LL (1983): Radiation of human mitochondrial DNA types analyzed by restriction endonuclease cleavage patterns. J Mol Evol 19:255–271.
Kidd KK, Kidd JR, Rogers J, Bowcock AM, Hebert JM, Lin A, Mountain JM, Cavalli-Sforza LL (1989): A study of over 100 markers in six populations favours a primary split between Africa and the rest of the world. J Cell Biochem Suppl 13C:119.
Kirch PV (1984): "The Evolution of Polynesian Chiefdoms." New York: Cambridge Univ Press.
Kocher TD, Wilson AC (1989): Mitochondrial DNA and human evolution. Presentation at ICN/UCLA colloquium, "Molecular Evolution," 27 Feb–6 March, Lake Tahoe, CA.
Kocher TD, Thomas KD, Meyer A, Edwards SV, Pääbo S, Villablanca FX, Wilson AC (1989): Dynamics of mitochondrial DNA evolution in animals: Amplification and sequencing with conserved primers. Proc Natl Acad Sci USA 86:6196–6200.
Mayr E (1950): Taxonomic categories in fossil hominids. Cold Spring Harbor Symp Quant Biol 15:109–117.
Mellars P (1989): Major issues in the emergence of modern humans. Curr Anthro 30(3):349–385.

Moritz C, Dowling TE, Brown WM (1987): Evolution of animal mitochondrial DNA: Relevance for population biology and systematics. Ann Rev Ecol Syst 18:269–292.

Nei M, Roychoudhury AK (1982): Genetic relationship and evolution of human races. Evol Biol 14:1–60.

Pääbo S (1989): Ancient DNA: Extraction, characterization, molecular cloning, and enzymatic amplification. Proc Natl Acad Sci USA 86:1939–1943.

Saiki R, Scharf S, Faloona F, Mullis KB, Horn GT, Erlich HA, Arnheim N (1985): Enzymatic amplification of beta-globin genomic sequences and restriction site analysis for diagnosis of sickle-cell anemia. Science 230:1350–1354.

Saitou N, Omoto K (1987): Time and place of human origins from mtDNA data. Nature 327:288.

Sibley CG, Ahlquist JE (1984): The phylogeny of the hominoid primates, as indicated by DNA-DNA hybridization. J Mol Evol 20:2–15.

Stoneking M, Cann RL (1989): African origin of human mitochondrial DNA. In Mellars PA, Stringer CB (eds): "The Human Revolution." Edinburgh: Edinburgh Univ Press, pp 17–30.

Stoneking M, Bhatia K, Wilson AC (1986): Rate of sequence divergence estimated from restriction maps of mitochondrial DNAs from Papua New Guinea. Cold Spring Harbor Symp Quant Biol 51:433–439.

Stringer CB, Andrews P (1988): Genetic and fossil evidence for the origin of modern humans. Science 239:1263–1268.

Stringer CB (1990): Replacement, continuity, and the origin of *Homo sapiens.* In Bräuer G, Smith F (eds): "Continuity or complete Replacement? Controversies in the Evolution of *Homo sapiens.*" Balkema, Zagreb, in press.

Toth N, Schick KD (1986): The first million years: The archaeology of protohuman culture. In Schiffer M (ed): "Advances in Archaeological Method and Theory." London: London Academic Press, pp. 1–96.

Vigilant L, Pennington R, Harpending H, Kocher TD, Wilson AC (1989): Mitochondrial DNA sequences in single hairs from a Southern J African population. Proc Natl Acad Sci USA 86:9350–9354.

Wainscott J (1987): Out of the garden of Eden. Nature 325:13.

Wallace DC, Zheng X, Lott MT, Shoffner JM, Hodge JA, Kelly RI, Epstein CM, Hopkins LC (1988): Familial mitochondrial encephalomyopathy (MERRF): Genetic, pathophysiological, and biochemical characterization of a mitochondrial DNA disease. Cell 55:601–610.

White TD, Johanson DC, Kimbel WH (1981): *Australopithecus africanus:* Its phylogenetic position reconsidered. South African J Science 77:445–470.

Wilson AC, Cann RL, Carr SM, George M, Gyllensten U, Helm-Bychowski KM, Higuchi RG, Palumbi SR, Prager EM, Sage RD, Stoneking M (1985): Mitochondrial DNA and two perspectives on evolutionary genetics. Biol J Linn Soc 26:375–400.

Wilson AC, Stoneking M, Cann RL, Prager EM, Ferris SD, Wrischnik LA, Higuchi RG (1987): Mitochondrial clans and the age of our common mother. In Vogel F, Sperling K (eds): "Human Genetics," Berlin: Springer-Verlag, pp 158–164.

Wolpoff MH (1989): Multiregional evolution: The fossil alternative to Eden. In Mellars PA, Stringer CB (eds): "The Human Revolution." Edinburgh: Edinburgh Univ Press, pp 62–108.

Wyles JS, Kunkel JG, Wilson AC 1983): Birds, behavior, and anatomical evolution. Proc Natl Acad Sci USA 80:4394–4397.

New Perspectives on Evolution, pages 225-250

Molecular Biology and Evolutionary Theory: The Giant Panda's Closest Relatives

STEPHEN J. O'BRIEN, RAOUL E. BENVENISTE, WILLIAM G. NASH, JANICE S. MARTENSON, MARY A. EICHELBERGER, DAVID E. WILDT, MITCHELL BUSH, ROBERT K. WAYNE, AND DAVID GOLDMAN

Laboratory of Viral Carcinogenesis, National Cancer Institute, Frederick Cancer Research and Development Center, Frederick, Maryland 21702-1201 (S.J.O., R.E.B., W.G.N., J.S.M., M.A.E., D.E.W.); H & W Cytogenetic Services, Inc., Lovettsville, Virginia 22080 (W.G.N.); National Zoological Park, Smithsonian Institution, Washington, DC 20008 (D.E.W., M.B.); Laboratory of Clinical Studies, National Institute of Alcohol Abuse and Alcoholism, National Institutes of Health, Bethesda, Maryland 20892 (D.G.); Department of Biology, University of California at Los Angeles, Los Angeles, California 90024 (R.K.W.).

INTRODUCTION

The field of molecular evolution is over twenty-five years old. The discipline is comprised of two distinct components: The collection of molecular data which can be used for phylogenetic inference and the development of evolutionary algorithms for interpreting these data. Both areas have matured substantially and as a result the use of molecular data is today considered a critical element of most evolutionary discussions. We have used a variety of molecular technologies (DNA hybridization, genetic distance using homologous proteins, immunological distance, mitochondrial DNA maps, and cytological/syntenic homology) to study genome evolution in mammals. These studies have permitted the development of consensus molecular phylogenies of several carnivore families including ursids, felids and canids. In this article we illustrate the rationale and methods of molecular evolution as they have been applied to the evolution and systematics of ursids (bear family), the giant and red pandas, and the raccoon family of Carnivora. A general agreement of different molecular methods for the same group provides credence to both the

This article is the text of an address which was also presented at a symposium entitled "Evolutionary Biology at the Crossroads" at Queens College in April 16 and 17, 1988. The article is printed in those proceedings as well by Queens College University Press (M. Hecht, editor) and is reproduced here with permission.

derived evolutionary trees, as well as to the efficacy of the methods for evolutionary inference.

A primary goal of evolutionary biology is the correct reconstruction of phylogenetic history, which would establish ancestral relationships between living and extinct species. This exercise can be conceptually divided into three components, namely (1) the derivation of an evolutionary tree which relates divergence nodes for ancient splits between evolutionary lineages, (2) the calibration of the tree with elapsed time for each of the divergence nodes, and (3) the interpretation of existing form and function of species in the context of evolutionary divergence and perhaps adaptation to a particular niche. Evolution and the sister discipline of taxonomy have remained largely in the purview of morphologists, ethologists, and physiologists since formal descriptive biology began. In 1962, E. Zuckerkandl and Linus Pauling introduced a new approach, one which studies the evolving genes and their protein products directly (Zuckerkandl and Pauling 1962). Since then the field of molecular evolution has made a number of insightful advances, but has also raised some fascinating questions about the tempo and mode of molecular and morphological evolution.

The application of molecular data to the questions of evolutionary divergence is made possible by the emergence and application of the molecular clock hypothesis, a simple yet very powerful concept. The hypothesis is based on the premise that the genetic material (DNA) of reproductively isolated species diverges continuously over time because of random mutations in chromosomal DNA which are passed on to subsequent generations. These mutations often occur in the coding regions of DNA that determine the amino acid sequence of a protein. So mutational changes between two related species can be studied by examining either DNA or the protein products of a gene sequence.

As species evolve, these substitutions continue to accumulate and with increased time, the extent of sequence divergence becomes greater. The mutational differences are mostly evolutionary "noise," but have the advantage of being proportional to the time elapsed since the existence of a common ancestor of two species. Further, for any individual molecular metric, the time of divergence can be calibrated by measuring the same metric between species whose time of divergence has been established geologically (for example, the time of separation of Old and New World occurred approximately 25 to 35 million years ago). Thus, by measuring one or more molecular metrics between two taxa, a relative estimate of their evolutionary distance and their time of divergence can be estimated. The use of the molecular clock has been elegantly applied to primate radiations and other evolutionary metrics by V. Sarich, A. Wilson, C. Sibley, and their colleagues and now is an established method of evolutionary theory (Sarich and Wilson 1967a, 1967b; Sibley and Ahlquist 1984, 1987; Wilson et al. 1977). The reader is referred to Wilson et al. (1977)

and Thorpe (1982) for a technical discussion of the molecular clock hypothesis and to Gribbin and Cherfas (1982) for an excellent popular description of the contributions of molecular evolution to sorting out the evolutionary history of mankind.

Molecular data and the associated molecular clock hypothesis have not been applied to evolutionary questions without controversy. Critics of the methods argue that different DNA sequences evolve at very different rates in the same lineage and that homologous DNA sequences often evolve at different rates in different lineages (Thorpe 1982). This leads to a perception of a clock which is sometimes uneven, misses a beat occasionally, and is not such a terrific timekeeper. It turns out that this flaw is not fatal because it is possible to check the rate of sequence evolution by a "relative rate test" (Sarich and Wilson 1967a, Wilson et al. 1977). Perhaps more importantly, the rate differential will affect limb lengths, but probably not the topology (position of the nodes) in an evolutionary tree. So although the molecular clock is not perfect, even the harshest critics agree that the amount of potentially valuable information present in the genetic material is enormous. For example, mammals contain about 3.2 billion evolving nucleotide base pairs in a linear array in their germ cells. Furthermore, the critics might also agree that mutational accumulation is cumulative and proportionate to the time elapsed since two measured species shared a common ancestor.

A variety of molecular methods have been used for phylogenetic inference in the past 25 years (O'Brien et al. 1985, Nei 1987, Ayala 1976). In parallel with the derivation of comparative molecular data, a number of mathematical algorithms have been developed which are useful for deriving topologies from genetic distance as well as from unweighted unit character data sets (Fitch and Margoliash 1967, Farris 1972, Fitch 1981, Dayhoff 1976, Felsenstein 1984, Swofford 1985, Sneath and Sokal 1973). We reasoned that the multifaceted molecular approach would be beneficial to a long-standing evolutionary puzzle, the phylogeny of the giant panda, the red panda, and the bears. First, the strategy would provide new data to a question which had been debated for a century. Secondly, it would allow us to compare the data sets derived from several molecular methods for the same group to each other and to similar data sets from a related mammalian group, anthropoid apes and man. Thirdly, if a verifiable consensus relationship were derived, then knowledge of this would allow a retrospective evaluation of the molecular methods as well as the morphological characters which led to alternative interpretations in the past. The results of our analysis, which we will review here, were strikingly concordant in their predicted topologies (O'Brien et al. 1985, O'Brien 1987, Goldman et al. 1989). The only ambiguity which the data could not resolve was the relationship among four recently evolved species of ursine bears. The results which are presented reveal several noticeable limitations (and strengths) about the application of these methods to phylogenetic inference.

BACKGROUND

The giant panda, *Ailuropoda melanoleuca,* was first described to the West by the French missionary-naturalist Pere Armand David in 1869. David recognized the giant panda as a scientific novelty and he dispatched a description of this new species, which he termed *Ursus melanoleuca* (which means black and white bear), to his colleague Alphonse Milne-Edwards, son (and later successor) of the director of the Paris Museum of Natural History (David 1869). The following year, after examining skins and skeletal material David sent, Milne-Edwards decided that several osteological and dental characteristics were not typically ursid (bear-like) but more closely resembled the giant panda's smaller Chinese cousin, the lesser or red panda, *Ailurus fulgens.* Since these and other characters had placed the red panda in the raccoon family, Procyonidae, Milne-Edwards concluded that the giant panda was really a giant procyonid which had developed certain bear-like traits by evolutionary convergence (Milne-Edwards 1870). This scientific exchange marked the beginning of a century-old debate regarding the correct phylogenetic positions of both the giant and red pandas (Table I).

Although the giant panda certainly looks like a bear, it has some unique characteristics and habits that are unusual for bears. First, the giant panda, like the red panda, is largely herbivorous, subsisting on a diet of bamboo shoots, stems and leaves. Their bamboo diet has led to some specialized or derived morphological adaptations, some of which are apparent in the panda's Miocene ($\geq$ 10 million years old) progenitors. The giant panda's teeth are large with low flat cusps; its skull and jaw are massive with enlarged jaw muscles for extra grinding power. The giant panda has relatively massive forequarters which account for its ambling gait. The animal has a sixth digit on the forepaw resulting from an evolutionary extension of the radial sesamoid wrist bone to an awkward but functional opposable thumb. The combination of these characters leads one to the conclusion that the giant panda is specialized for sitting on its hind quarters for long periods of time eating bamboo.

The giant panda has other features that are atypical for bears but that are not related to its diet. The male genitalia are tiny, and posteriorly directed in a manner similar to the procyonids. The panda does not really behave like a bear either. Certain alpine bear species exhibit a type of annual hibernation, whereas the panda does not. This may also be related to diet because pandas cannot store enough energy from bamboo, a relatively inefficient energy source. Ethologists have used vocalizations to demonstrate phylogenetic relationships; here again, the panda and bears differ. Bears roar or growl, while the giant panda bleats, a rather unbear-like vocalization which is more reminiscent of sheep or goats.

The history of phylogenetic and taxonomic assignments of the giant and red pandas has been less than reassuring. Over the last century, there have been

TABLE I. Taxonomic Assignments of the Pandas

Date	Giant panda	Red panda	Basis	Author
1869	Ursidae	—	Morphology	P. A. David
1870, 1874	Procyonidae	Procyonidae	Osteological characters and dentition similar to lesser panda	H. Milne-Edwards
1870, 1875	Ursidae	—	Intracranial cast; skeletal morphology	P. Gervais
1885	Procyonidae	Procyonidae	Skull architecture, dental morphology	St.-G. Mivart
1891	Ursidae	Procyonidae	Review of mammals similarity to fossil ursid, Hyaenarctos	W. H. Flower
1895	Ursidae	Procyonidae	Skeletal morphology similar to fossil ursid Hyaenarctos	H. Winge
1901	Procyonidae	Procyonidae	Skeletal and dental morphology	E. R. Lankester
1901	Procyonidae	Procyonidae	Skull and limb morphology	R. Lydekker
1902	Ursidae	Procyonidae	Review of mammals	F. E. Bedderd
1904	Ursidae	Procyonidae	Similarity to fossil ursid Hyaenarctos	M. Webber
1913	Ursidae	—	Dental and osteological morphology	K. S. Bardenfleht
1921, 1928	Ailuropodidae	Ailuridae	Review of procyonid taxonomy	R. I. Pocock
1936	Procyonidae	Procyonidae	Skull and dental morphology	W. K. Gregory
1936	Procyonidae	Procyonidae	Visceral anatomy	H. C. Raven
1943	Ursidae	Ursidae	Morphology of auditory region and ossicles	W. Segall
1945	Procyonidae	Procyonidae	Review of mammals	G. G. Simpson
1946	Ursidae	—	Brain topology	F. A. Mettler and L. J. Gross
1951	Procyonidae	Procyonidae	Review of mammals	J. R. Ellerman and T. C. S. Morrison-Scott
1955	Procyonidae	—	Review of vertebrates	E. Colbert
1956	Ursidae	—	Precipitin test, serum proteins	C. A. Leone and A. L. Wiens
1964	Procyonidae	—	Review of mammals	E. P. Walker
1964	Ursidae	Procyonidae	Comparative anatomy	D. D. Davis
1966	Procyonidae	Procyonidae	Ethological characters	D. Morris
1964, 1966	Procyonidae	Procyonidae	Karyology, cytology	R. E. Newnham and W. M. Davidson

(continued)

TABLE I. Taxonomic Assignments of the Pandas (continued)

Date	Giant panda	Red panda	Basis	Author
1971	Ailuropodidae	—	Anatomy	N. Kretozoi
1973	Ursidae	Ursidae	Immunological distance	V. Sarich
1973	Procyonidae	Procyonidae	Review of Carnivores	R. F. Ewer
1974	Ailuropodidae	—	Morphology, dentition, ethology	C. Chu
1974	Ailuropodidae	—	Fossil evidence, cranial and dental morphology	T. K. Wang
1978	Ursidae	—	Review	D. Starck
1979	Ailuropodidae	Ailuridae	Review of paleontological, morphological, serological, karyological and ethological characters of giant panda	V. E. Thenius
1980	Ailuropodidae	Procyonidae	Banded karyology	D. Wurster-Hill and M. Bush
1980	Ursidae	—	Dentition, serology	Q. B. Hendey
1981	Ursidae	—	Serology	W. Pan, L. Chen, and N. Xiao
1981	Ailuropodidae	Ailuridae	Review of mammals	J. F. Eisenberg and H. Setzer
1983	Ursidae	Procyonidae	Review of mammals	R. M. Nowak and J. L. Paradiso
1984	Procyonidae	Procyonidae	Review of mammals	B. Bertram
1985	Ailuridae	Ailuridae	Behavior, reproduction	G. Schaller
1985	Ursidae	Procyonidae	Ridges on the hard palate	M. Eisentraut
1985	Ursidae	—	Globin sequence	G. Braunitzer et al.
1986	Ailuropodidae	Ailuridae	Globin sequence	D. A. Tagle et al.
1986	Ailuropodidae	—	Comparative anatomy	Consortium Beijing Zoo Beijing Agricultural University Beijing Natural History Museum

Table I presents a summary of the conclusions of the taxonomic and systematic treatises relating to the giant panda published since 1869. Of 42 citations, 18 supported placement of the giant panda in Ursidae, 13 in Procyonidae (raccoon family), and 11 in Ailuropodidae or Ailurinae (separate family). Specific references can be found in (Davis 1964, Morris and Morris 1981, Thenius 1979, Jarofke and Ratsch 1985). Edwin Colbert interpreted the controversy in 1938 (Colbert 1938) as follows.

"So the question has stood for many years, with the bear proponents and the raccoon adherents and the middle-of-the-road group advancing their several arguments with the clearest of logic, while in the mean time the giant panda lives serenely in the mountains of Szechuan with never a thought about the zoological controversies he is causing by just being himself."

over 40 different systematic treatises directed toward the two pandas, some with new information, but many which reinterpreted previous data. With almost equal frequency, it has been concluded that the giant panda is (1) a specialized bear, Ursidae; (2) a specialized member of Procyonidae; (3) neither, but constitutes a separate Carnivore family, Ailuropodidae, comprised of only the giant pandas, or shared with the red panda (see Table I).

The systematic arguments were multifaceted; the data were apparently contradictory, and the debates were often aligned in national or cultural camps. In the same year that Milne-Edwards opted for procyonids, P. Gervais examined the brain morphology and joined the bear camp (Gervais 1870). In 1885, George St. Mivart reviewed the classification of arctoid (bear and dog-like) carnivores and placed the giant panda in Procyonidae because he found its similarities to the red panda inescapable (St. Mivart 1885). St. Mivart's conclusions were reaffirmed by an impressive collection of British and American naturalists over the next century. The prolific British taxonomist, R. I. Pocock, was the first to suggest that the giant panda deserved separate ranking and thereby raised the panda to its own family status, Ailuropodidae (Pocock 1921).

In 1964, D. Dwight Davis, then curator of mammals at Chicago's Field Museum of Natural History, published an extraordinary monograph based upon the gross anatomy of Su Li, a captive male panda who died in 1938 at the Brookfield Zoo (Davis 1964). Davis's opus, which Stephen Jay Gould recently characterized as "our century's greatest work of comparative anatomy" (Gould 1986) neatly described some 50 organ systems. His taxonomic conclusions were resounding: ". . . the giant panda is a bear and . . . very few genetic mechanisms—perhaps no more than half a dozen—were involved in the primary adaptive shift from *Ursus* to *Ailuropoda*" (Davis 1964). Davis's conclusions were quickly accepted by such authorities as Gould and Ernst Mayr who considered the matter finally settled (Gould 1986, Mayr 1986). But others did not agree, and with good reason. Davis's analysis, although lengthy, was largely anatomical, but did not follow standard principles of systematics. His critics argued that many of the traits he marshaled were irrelevant for phylogenetic concerns, since traits shared by bears and pandas were found in many other carnivores as well (MacIntyre and Koopman 1967, Morris and Morris 1981, Ewer 1973). By his own admission, Davis simply concluded a priori that the giant panda was a bear and assumed this throughout his text, making no attempt to present comparative data because ". . . this became so difficult, I [Davis] gave up." Gould recently remarked that "Davis's personal tragedy must reside in his failure to persuade his colleagues" (Gould 1986).

R. F. Ewer, in her excellent monograph on Carnivora (Ewer 1973), agreed with a comprehensive review of the panda's ancestry published in 1966 by Ramona and Desmond Morris, mammal curators at the London Zoo (Morris and Morris 1981). Based on both behavior and morphology the Morrises concluded that the giant panda was a procyonid. More recently, ethologists

John Eisenberg (Eisenberg et al. 1973) and George Schaller and colleagues (Schaller et al. 1985) independently argued for separate family status as did several paleontologists (C. Chu, T. Wang, W. C. Pei, E. Thenius) who examined the meager fossil record of ursids and procyonids (Chu 1974, Wang 1974, Pei 1974, Thenius 1979). The first banded chromosomal study lent support to that view, since pandas have 42 largely metacentric (biarmed) chromosomes while most bears have 74 largely acrocentric (single armed) chromosomes, a dramatic difference (Wurster-Hill and Bush 1980).

Finally, in 1986 a text on the anatomy of the giant panda which was just as daunting as Davis's was published in China (Beijing Zoo 1986). A consortium of zoologists from the Beijing Zoological Gardens and the associated universities presented 600 pages of anatomical data based on 27 specimens. They concluded that "the giant panda is different from the bear. . . . [We] are in favor of assigning an independent family for the giant panda." Suffice it to say that 120 years after its first description to Western naturalists, the panda's origins remain controversial.

In retrospect, it seems like this controversy was hardly a trivial issue, but one of meaningful consequences to a wider understanding of evolutionary processes. How much morphological or molecular change has occurred in this lineage? What were the driving forces of speciation, of morphological divergence? How widely divergent were the panda's or bears' ancestors in the distant past? The giant panda may not care whether it is a bear, a raccoon, or a solitary member of an ancient family (see legend to Table I), but our ability to perceive and interpret biological changes that occurred in the past are diminished without this kind of understanding.

The dilemma of the panda's ancestry also provided an excellent opportunity to examine how effective different types of biological characters are in providing information about how organisms are related. The reason is that the methods and measures of phylogenetic inference come in two packages. The first (called homologous traits) are similarities between taxa that are related by descent; that is, characters that were inherited from a recent common ancestor. Evolutionary relatedness is, in large part, measured by counting up the number of shared homologous characteristics which occur in different species. The second kind of characters (termed *analogous traits*) owe their similarity to evolutionary convergence. These traits are morphological or behavioral similarities that may be a selected response to a common environmental condition (e.g., prey base, predators, pathogens). An example of convergence is the development of flying ability in birds, bats and insects. Because analogous (convergent) characters have no basis in common descent, they actually confound the deductive process and mislead the naturalist. The confusion between homology and analogy represents the crux of the paradox which fueled the earlier disputes.

A common difficulty in identifying homologous morphological and anatomical traits is that their genetic basis is rarely well understood. For example, a small change in a morphological trait may involve large amounts of reorganization at the gene level, while other more complex traits (e.g., like the development of the panda's thumb) may actually involve a rather small amount of genetic change. Dwight Davis had suggested exactly this, but his conclusions were speculative since the accurate weighing of characters without a demonstrated genetic basis is virtually impossible.

THE MOLECULAR METHODS

In order to study the DNA and encoded proteins of the pandas and their relatives, we collected blood samples and small skin biopsies from all the species listed in Table II. Blood samples were separated by centrifugation into erythrocytes, leukocytes and plasma. The skin biopsy was digested with trypsin plus collagenase and used to establish a primary tissue culture line (Modi 1987). DNA and/or soluble proteins were extracted from each of the materials and employed in the methods listed in Table III. Each method produced a measure of "evolutionary distance," a quantitative estimate of DNA sequence divergence between measured species. Matrices of these distance values were used to construct phenetic topologies using evolutionary algorithms presented in Table IV. In addition the electrophoretic data (allozyme and 2-dimensional gel electrophoresis, 2DE) were also treated as unweighted unit-character data sets. In this treatment, each electrophoretic form (allele or fixed difference between species) is considered as a discrete character which was present or absent. The character matrix was analyzed using the Wagner method contained in the PAUP (phylogenetic analysis using parsimony) computer package of Dr. D. Swofford (Swofford 1985). This algorithm derives unrooted minimum length phylogenetic trees based on the principle of parsimony. Roots of trees were identified as the midpoint between extreme taxa, in this case bears and raccoons.

The first molecular method we used was DNA hybridization of unique cellular DNA (Fig. 1), a commonly used procedure for measurement of phylogenetic affinity (Sibley and Ahlquist 1984, 1987; O'Brien et al. 1985; Benveniste and Todaro 1976). Briefly, DNA from one species is labeled by growing skin fibroblast cells in medium containing radioactive DNA precursors. DNA from a second related species is unlabeled or "cold". Both DNAs are extracted and sheared by sonication to pieces of about 400 nucleotides in length, and heated to "melt" the original double-stranded DNA hybrids to single-strand fragments. A mixture of the two melted DNAs is allowed to reanneal. The amount of hybrid DNA molecules is measured by collecting radioactive hybrid molecules on a filter or by chromatography. Two measures of sequence homology are obtained from this procedure: First, the percent hybridization between

TABLE II. Species and Sources

Code	Species	Common name	Source	Sex	Director	Veterinarian
AME	*Ailuropoda melanoleuca*	Giant panda	National Zoo Wash., DC	M	T. Reed	M. Bush
AFU	*Ailurus fulgens*	Red panda	National Zoo Wash., DC	M	T. Reed	M. Bush
TOR	*Tremarctos ornatus*	Spectacled bear	Baltimore Zoo Baltimore, MD	F	B. Rutledge	M. Cranfield
UAR	*Ursus arctos*	Brown bear	Shrine Circus Omaha, NE	M	—	L. Philips
UAM	*Ursus americanus*	Black bear	Minnesota Zoo Minneapolis, MN	F	K. Roberts	F. Wright
UTH	*Ursus (Selenarctos) thibetanus*	Asiatic black bear	San Antonio Zoo San Antonio, TX	F	L. DeSabio	K. Fletcher
UMAL	*Ursus (Helarctos) malayanus*	Malayan sun bear	Henry Doorly Zoo Omaha, NE	M	L. Simmons	L. Philips
UMAR	*Ursus (Thalarctos) maritinus*	Polar bear	National Zoo Wash., DC	M	M. Robinson	M. Bush
UUR	*Ursus (Melursus) ursinus*	Sloth bear	National Zoo Wash., DC	F	M. Robinson	M. Bush
PLO	*Procyon lotor*	Raccoon	CRC, Front Royal National Zoo	M	M. Robinson	M. Bush

TABLE III. Metric of Evolutionary Distance Employed

Procedure	Basis	Data set
I. DNA hybridization	Melting curve, ΔT_m	Distance matrix
II. Genetic distance (Nei) isozymes	N = 50 isozymes	Distance matrix and cladistics
III. Genetic Distance 2D gel fibroblast proteins	N = 261 proteins	Distance matrix and cladistics
IV. Immunological Distance	Albumin, Transferrin	Distance matrix
V. Karyology	G-trypsin banded high resolution 800 band stage	—

TABLE IV. Phylogenetic Algorithms

Algorithm	Computer program	Reference
1. UPGMA	Biosys PHYLIP	Sneath and Sokol 1973
2. MATTOP	MATTOP	Dayhoff 1976
3. FITCH-MARGOLIASH	PHYLIP	Fitch and Margoliash 1967, Felsenstein 1984
4. DISTANCE WAGNER	PAUP	Farris 1972, Swofford 1985
5. NEIGHBORLINESS	NEIGHBORLINESS	Fitch 1981

species A and species B; and second, the difference between the midpoints of the melting profile between heterologous DNA hybrids and that of homologous DNA. This latter measurement, called ΔT_m, is proportional to the amount of base-pair differences between the two DNA samples. Because all the single-copy cellular DNA is used (the repetitive DNA is removed before mixing) the procedure gives an estimate of the percent divergence of the over one billion base pairs of single-copy sequence in the mammalian genome.

The second molecular metric used was isozyme genetic distance (O'Brien et al. 1985, Nei 1972, 1978). In this procedure, extracts of soluble enzymes from different tissues (erythrocytes, leukocytes, fibroblast cultures) are separated by gel electrophoresis and stained histochemically (Fig. 2). The mobility of homologous enzymes from the different species (Fig. 3a) is compared and an estimate of the extent of mutational accumulation in the primary structure of up to 50 proteins is determined. M. Nei and others developed very useful mathematical formulae for estimating the extent of allelic mutational substitution at a group of loci between populations (Nei 1972, 1978). The genetic distance estimate, D, is defined as the average number of gene differences per locus between individuals in two test populations. Within the

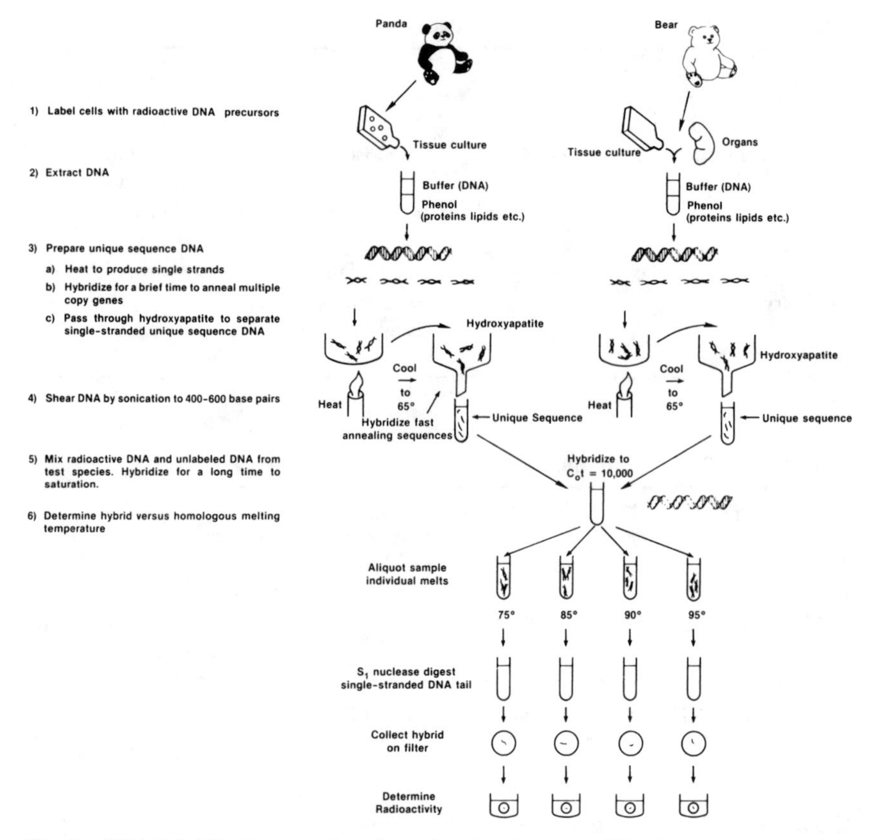

Fig. 1. DNA hybridization procedure for estimating thermal stability from a gradual melt of heterologous DNA hybrids. The validity of a data matrix derived by molecular methods like DNA hybridization can be tested in three ways (O'Brien et al. 1985). The first test is reciprocity; that is, ΔT_m values using species A as the labeled DNA and species B as the cold DNA should have a low variance from the values obtained when species B is labeled and species A is cold. The second test is the relative rate test. This test demands that a species which is outside the group being studied should have an equal molecular distance from all the species in the studied group. For example, we used a dog as an out-group and this showed an equivalent distance from bears, procyonids, and pandas. If that had not occurred, it would indicate that the molecular clock was not running on time in that lineage. The third test is based on the prediction that a data set which is an accurate reflection of evolutionary distance between taxa should obey the triangle inequality. That is, for distances between any three taxa in the matrix, one distance cannot exceed the sum of the remaining two. If this condition is met the data are said to be *metric.* Unfortunately, some distance estimates (e.g., Nei genetic distance) are non-metric for theoretical reasons and cannot be tested. Compliance of a molecular distance data set to these three tests is taken as evidence that the matrix will produce the correct evolutionary relationship.

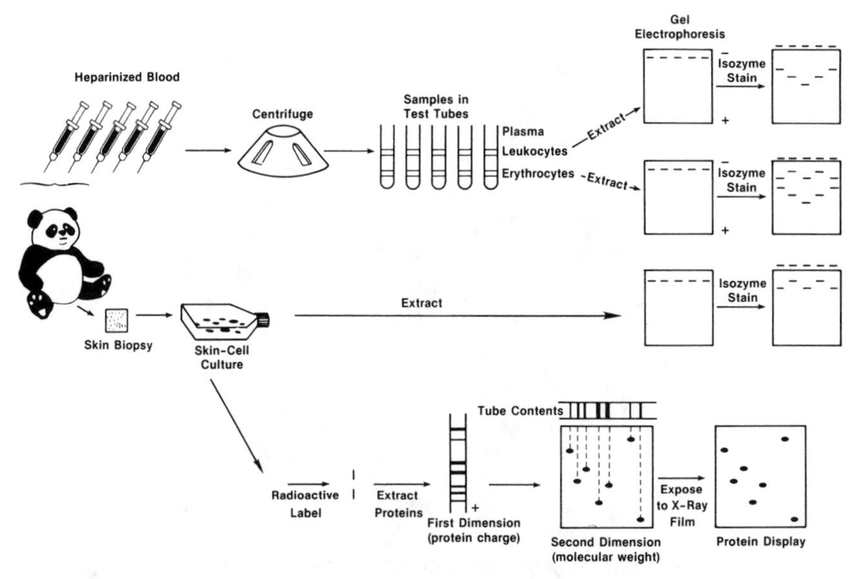

Fig. 2. Two electrophoretic methods for measuring genetic distance are illustrated. In one method, fresh blood is treated with heparin to prevent clotting so that different components (white cells, red cells, and plasma) can be separated by centrifugation. In addition, a skin biopsy from the animal is digested with trypsin and collagenase and used to establish a tissue culture cell line. Soluble enzymes from extracts of blood components and the cell lines are subjected to electrophoresis; that is, they are exposed to an electric field that causes them to migrate through a gel matrix, after which they are visualized by specific stains for enzyme reaction products. Enzymes that have accumulated a single amino acid difference migrate to a different position from those that have not. By comparing mobility at some 50 homologous isozyme systems between two species, an estimate of the frequency of identical gene products is achieved. The genetic distance is the negative logarithm of this value and increases proportionally with evolutionary time. Three tissues (red cells, white cells, and tissue culture cells) are used to resolve tissue specific enzymes and to provide duplications of equivocal results in one tissue. In the second method, radioactively labelled proteins from skin cell cultures are exposed to electric fields that separate them in two dimensions. Proteins are separated first on the basis of electric charge and then on the basis of molecular weight. Finally, the gels are exposed to X-ray film, which reveals hundreds of proteins whose position can be compared between different species. Genetic distance based on shifts of the presumed homologous protein gene products are computed exactly like the isozyme genetic distance and calibrated with evolutionary time.

limits of certain assumptions relating to the electrophoretic resolution and relative rates of amino acid substitution, the genetic distance values increase proportionately with the time the compared populations have been reproductively isolated.

Our third method was to compute genetic distance from a different class of cellular proteins, those resolved by 2DE (Goldman et al. 1989). Like the

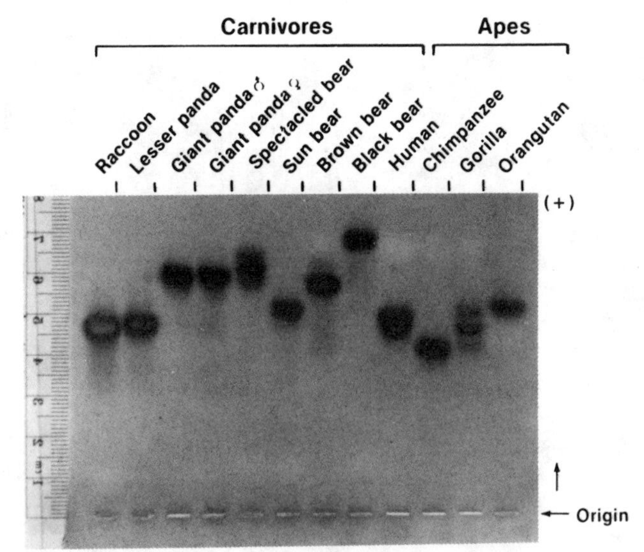
Carnivores
Apes
Raccoon
Lesser panda
Giant panda ♂
Giant panda ♀
Spectacled bear
Sun bear
Brown bear
Black bear
Human
Chimpanzee
Gorilla
Orangutan
(+)
Origin
a
PGD (6-phosphogluconate dehydrogenase)

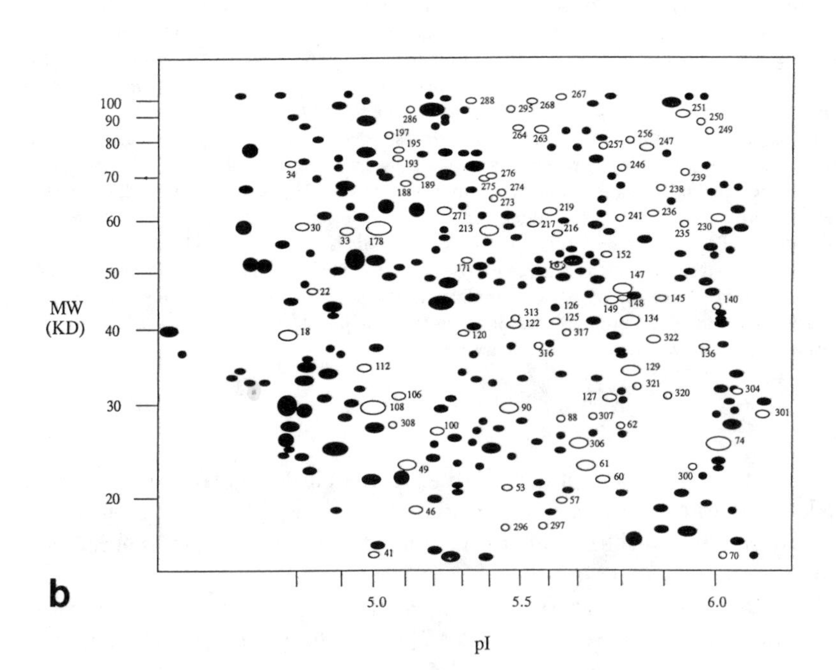
MW
(KD)
100
90
80
70
60
50
40
30
20
5.0
5.5
6.0
pI
b

isozyme procedures, this method measures the extent of protein migration on electrophoretic gels (Fig. 2), but it has the advantage of resolving over 300 radiolabeled fibroblast proteins on a single gel (Fig. 3b).

The fourth molecular method involved data collected and published previously in a classic paper by Vincent Sarich (1973). Sarich, who pioneered the use of immunological distance as a method for phylogenetic resolution, tested the giant panda and its relatives in 1972 (Sarich and Wilson 1967a, 1967b; Sarich 1973). The procedure he employed measures amino acid differences between homologous serum proteins (e.g., albumin) from two species based on displacement of immunological titration curves in a micro-complement fixation assay for antibody. Briefly, several rabbits were immunized with purified albumin from the giant panda. The sera was then pooled and titered against giant panda albumin. In an evolutionary distance determination, albumin from another species, e.g., a raccoon, is first preincubated with titered antiserum against the giant panda albumin. Then, the adsorbed serum is retested against giant panda albumin. The remaining antibodies bind to the giant panda albumin in an amount quantitatively related to the amount of immunological difference. The displacement of immunological titration curves is proportional to the sequence divergence between the albumin genes of the two species. When several antisera are prepared, a matrix of immunological distances can be used as above for estimating evolutionary distances. Sarich generated such immunological distance matrices with two proteins, albumin and transferrin (Sarich 1973).

THE SOLUTION

The results of each of our different evolutionary measurements have been published elsewhere (O'Brien et al. 1985, O'Brien 1987, Goldman et al. 1989) and will be summarized briefly here. The evolutionary distances derived from each method were by and large concordant in the topology which was produced. Different clustering or phylogenetic algorithms for the same data sets in general produced the same tree. Uncertainties within certain data sets were apparent regardless of the algorithm and were often resolved using other methods. The one area which was not resolved was the relationship among the

Fig. 3. **A:** Electrophoretic migration is shown for one isozyme, 6-phosphogluconate dehydrogenase, using extracts of tissue culture cells. The raccoon and red panda enzymes have identical mobilities, while the giant panda and the brown bear share the same mobility, which is distinct from the raccoon. The spectacled bear is a heterozygote; i.e., there are two genetic forms—one is the same as the giant panda and the other species. Both the sun bear and the black bear have unique electrophoretic forms not seen in the other species. The hominoid ape tissues were run in the same gels. **B:** Diagrammatic representation of 2DE gel patterns of bears, pandas, and procyonids. Open ellipses vary between species; closed ellipses represent invariant proteins (Goldman et al., 1989).

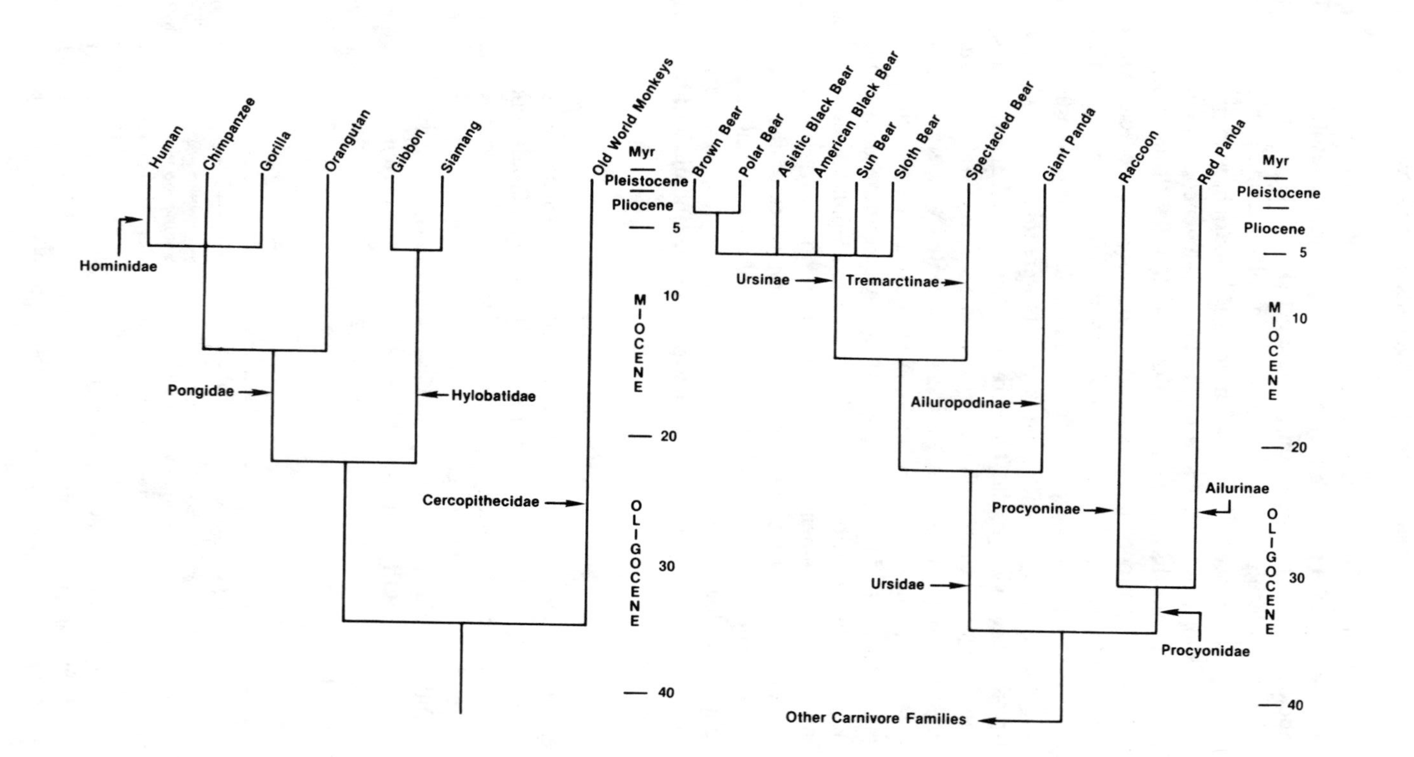

Human
Chimpanzee
Gorilla
Orangutan
Gibbon
Siamang
Old World Monkeys
Hominidae
Pongidae
Hylobatidae
Cercopithecidae
Myr
Pleistocene
Pliocene
5
10
MIOCENE
20
OLIGOCENE
30
40
Brown Bear
Polar Bear
Asiatic Black Bear
American Black Bear
Sun Bear
Sloth Bear
Spectacled Bear
Giant Panda
Raccoon
Red Panda
Ursinae
Tremarctinae
Ailuropodinae
Procyoninae
Ailurinae
Ursidae
Procyonidae
Other Carnivore Families
Myr
Pleistocene
Pliocene
5
10
MIOCENE
20
OLIGOCENE
30
40

ursine bears (genus *Ursus*) which diverge rather close in time from a common ancestor 6 to 10 million years before the present (mybp). A consensus tree based upon combination of results from each of the four molecular methods is presented in Figure 4.

Once we found that the four different molecular topologies were in agreement, our next task was to place a time scale on the evolutionary tree. This turned out to be a difficult problem. Although the carnivore fossil record is unusually good and has been studied intensively, even the commonly accepted dates for divergence of these families can vary by as much as 25 to 50 percent (Ewer 1973, Savage and Russell 1983, Eisenberg 1981, Flynn and Galiano). For example, geological dates for the time of procyonid-ursid divergence range from 30 to 50 million years ago. A strategy we have used for setting our molecular clock was to take advantage of the demonstration that the primate and carnivore clocks appear to run at the same rate (Benveniste 1985, Kohne et al. 1972, Sarich 1969). Because the primate radiations have been studied rather extensively, we reasoned that the species pairs that had the same distance for a particular metric within the two orders shared a common ancestor at approximately the same time in their evolutionary history. Further, that time in the carnivores might be precisely identified by the primate time scale that has been calibrated by numerous authors from both paleontological and molecular perspectives (Sarich and Wilson 1967a, 1967b; Sibley and Ahlquist 1984, 1987; Wilson et al. 1977; Thorpe 1982; Gribbin and Cherfas 1982).

We performed the same three molecular analyses on the hominoid primates and used Sarich and Wilson's data on the same group using immunological distance (Sarich and Wilson 1967a). Our primate results confirmed precisely what others have reported (Sarich and Wilson 1967a, 1967b; Sibley and Ahlquist 1984, 1987) so we could now compare primate and carnivore molecular phylogenies (presented in Fig. 4). For calibration of the primate phylogenies we used the time scale based on paleontologist Peter Andrews's estimated date of human-orangutan divergence at approximately 13 million years ago (Andrews 1986). The derived primate topology places the time of human-Old

Fig. 4. A consensus molecular phylogeny of the bear family, the giant and red pandas, and raccoon. This tree is a subjective compromise of the results of the following four different molecular methods: (1) DNA hybridization, (2) isozyme genetic distance, (3) 2D gel electrophoresis of fibroblast proteins, and (4) immunological distance. Each of these methods gave a similar evolutionary tree and they differed largely by length of the limbs. The primate tree was built using the same four methods: first, to demonstrate that our laboratory's molecular data gave us the same answers that other molecular laboratories were producing, and second, to provide an index of calibration dates. Because the molecules we have studied seem to evolve at rates which were similar in primates and carnivores (Benveniste 1985, Kohne et al. 1972, Sarich 1969), we simply aligned the molecular scale of the carnivores with the primates and used the time scale of primate evolution to date the carnivore divergence nodes.

World monkey divergence at 25 to 35 million years ago, which agrees with generally accepted paleontological dates of these events. Because the four seemingly independent molecular methods were highly consistent within the carnivore radiations, we integrated the results and drew a consensus molecular phylogeny which included the definitive aspects of each phylogeny (Fig. 4).

Briefly, between 30 and 50 million years ago, the progenitors of the modern ursids and procyonids split into two lineages. Within 10 million years of that event, the procyonid group split into Old World procyonids (represented today by the genus *Ailurus,* the red panda) and the New World procyonids (raccoons, coatis, olingos and kinkajous). The red panda and giant panda clearly do not share a common ancestor after the ursid-procyonid split, emphasizing that the morphological similarities of the pandas are probably the result of convergence or parallel retention of ancestral characters. At about the same time as the gibbons split from great apes (18 to 25 million years ago), the ancestors of the giant panda diverged from the ursid line. This event was nearly 20 million years after the initial divergence of the ursid and procyonid lines. At the time that the orangutan diverged from the African ape-human line (13 to 16 million years ago) the earliest true bear, *Tremarctos* (spectacled bear), split from the ursid line. The genus *Ursus* began its radiation into the ursine bears (brown, black, and sun bears) 6 to 10 million years ago.

THE CHROMOSOMES

Our final analysis of the giant panda's phylogeny concerned what we originally thought would be a conflicting character, chromosome morphology. The six ursine bears all have 74 single-armed (acrocentric) chromosomes with the centromere at one end, while the giant panda has 42 mostly biarmed (metacentric) chromosomes. Earlier cytologists had thought that this was an irresolvable difference which argued against any recent heritage of the bear and giant panda (Wurster-Hill and Bush 1980, Newman and Davidson 1966). Because we were suspicious of this conclusion, we took another look. One of us (W. G. Nash) had recently developed some special techniques for producing high resolution G-banding patterns in carnivore chromosomes by transforming all primary cultures with a retrovirus and subsequently synchronizing the cells (Modi et al. 1987). Just before metaphase, there is a period of maximum chromosome extension during which cells could be harvested to produce banded chromosomes of exquisite detail.

When karyotypes of the six ursine bears were examined, they were virtually identical (Nash and O'Brien 1987). The surprise came when we compared bear chromosomes to those from the giant panda. It turned out that most of the giant panda's chromosome arms had an identical banding pattern to individual ursine bear chromosomes (Fig. 5a) (O'Brien et al. 1985). Giant panda chromosomes were actually pairs of bear chromosomes attached head to head with common centromeres. So what looked like a major chromosome reorganization was actually quite a simple fusion of ancestral bear chromosomes in the evolutionary line leading to the giant panda.

A similar but independent chromosome fusion event also must have occurred in the line which led to the spectacled bear (Fig. 5b). The reason is that the spectacled bear karyotype is also composed of biarmed fusions of acrocentric chromosomes found in the bears, but none of the fusion combinations seen in the spectacled bear are the same as the combinations found in the giant panda (Nash and O'Brien 1987). Similar chromosomal reorganizations, termed Robertsonian translocations, often have been suggested to be important in establishment of reproductive isolation during speciation events. It is tempting to hypothesize that the extensive chromosomal fusion may have played an adaptive role in the evolution of both the panda and spectacled bear lineages.

In order to fit the chromosome pattern of the red panda into this picture, one should first understand that much of the carnivore order is characterized by a rather conservative karyotype composed of metacentric chromosomes. For example, each of the 37 species of cats have either 18 or 19 pairs of mostly biarmed chromosomes and 15 of these chromosomes are invariant throughout the Felidae family (Wurster-Hill and Gray 1973, 1975). Furthermore, most of these conserved Felidae chromosomes are also present in several other carnivore families (procyonids, mustelids, hyaenids) either intact or only slightly variant (Wurster-Hill and Centerwall 1982, Modi and O'Brien, 1988). For this reason, it is useful to consider that the continued transmission of a conservative carnivore karyotype is preserved more or less intact in several families. Canids and ursids are exceptions to the rule because they both experienced a dramatic chromosome fissioning early on in their evolution. We can still align the dog and bear chromosome arms to certain segments of the primitive carnivore karyotype, but none of the ancestral carnivore fusion combinations are seen in modern dogs or bears (Nash and O'Brien 1987, Wayne et al. 1987). In these families the chromosome arms were shuffled and in many cases (as in the giant panda) reattached in different ways.

The banded chromosomes of the red panda are very similar to the primitive carnivore karyotype which is also seen in the procyonids (Wurster-Hill and Gray 1973). Fourteen chromosomes of the red panda are strikingly homologous to chromosomes found in several procyonids, and ten of these are identical to conserved carnivore chromosomes. Only two of the red panda's banded chromosomes had recognizable counterparts in the bears or giant panda. The simplest interpretation of these karyological results seems to affirm the molecular topologies; that is, the red panda and the procyonids possess a shared primitive carnivore karyotype suggesting they diverged at an evolutionary mode apart from the bear-giant panda divergence. The ursid line apparently evolved in conjunction with a global chromosomal fissioning away from the carnivore karyotype, resulting in a shared derived karyotype organization in

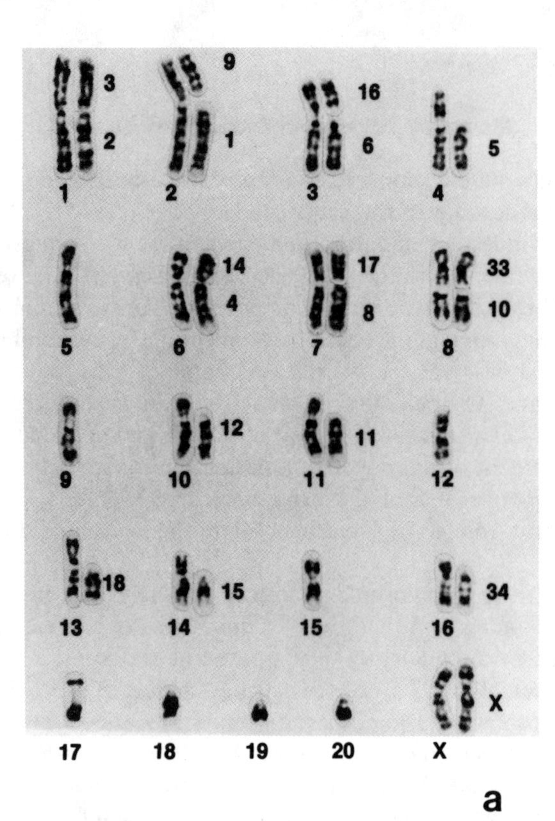

3 2 1
9 1 2
16 6 3
5 4
5
14 4 6
17 8 7
33 10 8
9
12 10
11 11
12
18 13
15 14
15
34 16
17 18 19 20 X X

a

7 2 1
8 4 2
9 11 3
13 10 4
21 14 5
25 1 6
20 5 7
27 12 8
26 16 9
30 18 10
31 11
32 12
33 13
34 14
36 15
3 16
6 17
17 18
15 19
24 20
35 21
23 22
22 23
29 24
28 19 25
X X

b

bears and giant panda. This chromosomal fissioning persisted in the ursine bears but was reorganized by centromere fusion twice: Once in the giant panda line and again after the split leading to the spectacled bear.

SOME IMPLICATIONS

The molecular data had told us that the giant panda's ancestors split off the ursid lineage 10 to 20 million years after the ursid-procyonid divergence, while the red panda split from the New World procyonids very near this time point. But how can we reconcile this conclusion with the many morphological and behavioral characteristics offered by several authors as evidence of an evolutionary distinction between bears and the giant panda? In retrospect, it is important to emphasize first that Dwight Davis not only stated, but meticulously detailed the fact that the vast majority of morphological characters actually affirmed the affinity of bears and the giant panda (Davis 1964). Second, many of the characters that giant pandas share with the red panda are related to the observation that both species are largely herbivorous (Morris and Morris 1981, Ewer 1973). Apparently, most of these traits (e.g., grinding teeth, massive skull, extra thumb, behavioral similarities) are really analogous or convergent—the misleading kind for phylogenetic inference. Both species apparently acquired the derived traits in parallel as an adaptative response to their diet. Other traits shared in the two pandas could simply be homologous primitive carnivore characters retained in pandas but arbitrarily lost in bears and New World procyonids.

An important consideration in solving phylogenetic puzzles is the information derived from the fossil record. There are about 50 described fossils of the giant panda dispersed throughout China and southeast Asia during the Pleistocene geological period (Chu 1974, Wang 1974, Pei 1974, Thenius 1979). Unfortunately, they are largely indistinguishable from modern pandas and tell us

Fig. 5. The recently evolved ursine bears (all except spectacled bears) all have 74 chromosomes with the centromere at the end (Nash and O'Brien 1987). The giant panda has 42 chromosomes, each with a centromere in the middle (O'Brien et al. 1985). **A:** When chromosomes of the brown bear *(Ursus arctos)* and giant panda *(Ailuropoda melanoleuca)* were compared, it became clear that nearly every large chromosome of the brown bear could be aligned with a giant panda chromosome arm. The left chromosome of each pair is the giant panda chromosome. The numbers below each pair refer to giant panda chromosomes. The numbers on the right of each pair identify brown bear chromosomes. Where no clear homolog exists with a bear chromosome, the panda chromosome is shown by itself. The giant panda seems to have developed its chromosome pattern by the relatively simple process of chromosome fusion at the centromere during its evolution off the main bear lineage. **B:** Comparison of G-banded chromosomes of spectacled bear, *Tremarctos ornatus* and *Ursus arctos.* The left chromosome of each pair is the spectacled bear chromosome. The numbers below each pair refer to spectacled bear chromosomes. The numbers to the right of each pair identify brown bear chromosomes.

little except that the panda's feeding adaptations were evident by 1 to 2 million years ago. The only earlier fossil which may be placed on the giant panda lineage with some certainty is a piece of mandible *(Agriarctos kretozo)* from a late Miocene site in Europe (Thenius 1979, Wayne et al. 1987). V. E. Thenius, who discovered the fossil, pointed to its incipient molarization of premolars as a precursor to the giant panda robust dentition. Thenius had suggested that the giant panda split from the ursid line 15 to 20 million years ago and that *Agriarctos* was on the panda line. Before that split, an early Miocene (about 20 to 25 million years ago) fossil genus *Ursavus* is thought to be the earliest ursid specimen. The placement of these fossils fits nicely with the molecular phylogeny we have derived.

The development of a consensus phylogeny of the giant and red pandas which is consistent from several biological perspectives now allowed us to make some taxonomic recommendations. The derived panda relationships indicated two possible options. They are (1) place the giant and red panda each in different families separate from each other and from Ursidae and Procyonidae, and (2) place the giant panda in Ursidae and the red panda in Procyonidae, preferably with subfamily status for both. The second option was chosen primarily because it acknowledges the relatively recent divergence of the giant panda from the ursid line and the definitive though somewhat older split of the red panda from the procyonid lineage. Molecular and fossil evidence had indicated that most carnivore families split from the main stock more than 40 million years ago (Wayne et al. 1989). Hence, the giant panda (which split 18 to 25 million years ago) is placed in Ailuropodinae, a subfamily of Ursidae, while the red panda is the single member of Ailurinae, a subfamily of Procyonidae (O'Brien et al. 1985).

A rather important outcome of the data here reviewed is the opportunity to compare directly the topology and rates of different molecular methods within the same group. We attempt this in Table V by presenting the approximate calibrated dates of five major divergence nodes as estimated from five molecular data distance matrices (O'Brien et al. 1985, Goldman et al. 1989, Sarich 1973). These values were obtained by making the three following presumptions: (1) molecular evolution in carnivores and primates proceeds at an equivalent rate (Benveniste 1985, Kohne et al. 1972, Sarich 1969); (2) the divergence date of orangutan from the ancestors of humans occurred about 13 myr ago (Andrews 1986); and (3) molecular evolution is proceeding in a fashion that is directly proportionate to elapsed time. The first presumption has been tested in several contexts and appears to be reasonably accurate. The human-orangutan divergence date is still under intense debate (for discussion see Sibley and Ahlquist 1987); however, the usual estimated range is 10 to 18 mybp. The comparative data presented in Table 5 belie the direct proportionality of all the molecular estimates with elapsed time since each data set gives substantive departure from precise quantitative agreement. For example, the

TABLE V. Estimated Date of Divergence Nodes in Ursid-Procyonid Taxa by Four Molecular Methods of Evolutionary Distance

			Genetic distance allozyme[a]		
Divergence node	AID	DNA Hybridization	2DE	1985	1988
Ursine bear radiation (Uam-Umar)	6.5	9.3	4.4	6.01	7.2
Spectacled bear-ursine	6.5	12.2	10.5	9.6	15
Giant panda-ursine	15	17.1	18.2	33.8	22
Lesser panda-raccoon	52	54.8	17.1	23.5	28
Procyonidae-ursidae	45	54.3	22.4	32.6	32.3

[a]Distances are expressed in myr before the present and are calibrated according to an estimated divergence of orangutan from other great apes of 13 mybp (see text). Allozyme results are independent data sets from O'Brien 1987, O'Brien et al. 1985. Data for DNA from O'Brien et al. 1985; AID from Sarich 1973.

dates of the ursine radiation node and the ursid-procyonid divergence node both differed by 100% between 2DE estimates and the DNA hybridization estimate (Table V). The divergence nodes in the middle of the topologies (closer to the calibration date of 13 myr) show better agreement between methods.

We interpret the lack of agreement in dating of divergence nodes despite excellent topological similarity to suggest that each method has a different relationship to elapsed time. For each method then, that relationship is not a simple direct proportionality between methods or with evolutionary time. For this reason, it seems important to employ several independent methods (molecular, morphological and paleontological) to reconstruct the correct phylogenetic interpretation of this or any other group.

While this study was progressing, it was difficult not to be impressed with the power of molecular and cytogenetic data, especially when multiple methods are employed, in resolving evolutionary relationships. The importance of this is evident in resolving evolutionary convergence, in conservation applications, and in unravelling the evolutionary significance of biological adaptation and speciation. The derivation of the panda's ancestral tree was not the only insight these studies produced. In addition, we discovered an outline of the relative divergence times of several other ursid and procyonid species, e.g., the giant panda results from an ancient split and need not be a punctuated event as the polar bear speciation seems to be (Goldman et al. 1989). We were able to compare various phenetic and phylogenetic algorithms for building trees and equally important to compare various molecular approaches to bear phylogeny—each approach with its own inherent biases and relationship with absolute time. By using the primate data matrices, we could compare the same molecular methods in that distantly related family. Finally, once a consistent molecular topology for any group is available, the interpretation of morphological form and function can be more readily achieved (Wayne et al., 1989).

REFERENCES

Andrews P (1986): Fossil evidence on human origins and dispersal. Cold Spring Harb Symp Quant Biol 51:419–428.

Ayala F (ed) (1976): "Molecular Evolution." Sunderland, MA: Sinauer Associates.

Beijing Zoo (Competent Authority), Beijing Univ, Beijing Agric Univ,

Beijing Sec Med Col, Beijing Nat Hist Mus, and Shaanxi Zool Inst (1986): "Morphology of the Giant Panda: Systematic Anatomy and Organ-Histology." Beijing: Science Press.

Benveniste RE (1985): The contributions of retroviruses to the study of mammalian evolution. In MacIntyre RJ (ed): "Molecular Evolutionary Genetics," Monographs Evol Biol Ser. New York: Plenum Press, pp 359–417.

Benveniste RE, Todaro GJ (1976): Evolution of type C viral genes: Evidence for an African origin of man. Nature 261:101–108.

Chu C (1974): On the systematic position of the giant panda *Ailuropoda melanoleuca* (David). Acta Zool Sinica 20:174–187.

Colbert EH (1938). The panda: A study in emigration. Nat Hist 42: 33–39.

David A (1869): Extrait d'une lettre du même, datée de la Principauté Thibetaine (independente) de Mou-pin, le 21 Mars. Nouv Arch Mus Hist Nat Paris, Bull 5:12–13.

Davis DD (1964): The giant panda: A morphological study of evolutionary mechanisms. Fieldiana Zool Mem 3. Chicago: Nat Hist Mus.

Dayhoff MO (1976): Survey of new data and computer methods of analysis: Atlas of protein sequence and structure. In Dayhoff MO (ed): "National Biomedical Research Foundation," Washington, DC: Natl Biomed Res Found, pp 1–8.

Eisenberg JF (1973): Cited Collins LR, Page JK (eds): "Ling-Ling and Hsing-Hsing, Year of the Panda." Garden City: Anchor Press/Doubleday.

Eisenberg JF (1981): "The Mammalian Radiations: An Analysis of Trends in Evolution, Adaptation and Behavior." Chicago: Univ Chicago Press.

Ewer RF (1973): "The Carnivores." New York: Cornell Univ Press.

Farris JS (1972): Estimating phylogenetic trees from distance matrices. Am Naturalist 106:645–668.

Felsenstein J (1984): Distance methods for inferring phylogenies: A justification. Evol 38:16–24.

Fitch WM (1981): A non-sequential method for constructing trees and hierarchical classifications. J Mol Evol 18:30–37.

Fitch WM, Margoliash E (1967): Construction of phylogenetic trees. Science 155:279–284.

Flynn JM, Galiano H. (year): Phylogeny of early Tertiary Carnivora, with a description of a new species of *Protictis* from the middle Eocene of Northwestern Wyoming. Am Mus Nov No. 2632:1–16.

Gervais P (1870): Mémoire sur les formes cérébrales propres aus carnivores vivants et fossiles. Nouv Arch Mus Hist Nat Paris 1 (6):103–162.

Goldman D, Rathna Giri P, O'Brien SJ (1989): Molecular genetic distance estimates among the Ursidae as indicated by one- and two- dimensional protein electrophoresis. Evol, 43:282–295.

Gould SJ (1986): Fuzzy wuzzy was a bear, Andy panda too. Discover (Feb):40–48.

Gribbin J, Cherfas J (1982): "The Monkey Puzzle: Reshaping the Evolutionary Tree." New York: Pantheon Books.

Jarofke D, Ratsch H (1985): A bibliography of the giant panda (*Ailuropoda melanoleuca*). In Klos H-G, Fradrich H (eds): "Proceedings of the International Symposium on the Giant Panda." Berlin: Bongo, vol 10, pp 209–228.

Kohne DE, Chiscon JA, Hoyer BH (1972): Evolution of primate DNA sequences. J Hum Evol 1:627–644.

Kurten B (1986): Reply to "A molecular solution to the riddle of the giant panda's phylogeny." Nature 318:487.

MacIntyre G, Koopman K (1967): Book review: "The Giant Panda. A Morphological Study of Evolutionary Mechanisms." Quart Rev Biol 42:72–73.

Mayr E (1986): Uncertainty in science: Is the giant panda a bear or a raccoon? Nature 323:769–771.

Milne-Edwards A (1870): Note sur quelques mammifères du Thibet oriental. Ann Sci Natl Zoo, Ser 5, art 10.

Modi WS, O'Brien SJ (1988): Quantitative cladistic analyses of chromosomal banding data among species in three orders of mammals: Hominoid primates, felids and arvicolid rodents. In Gustafson P, Appels (eds): "Stadler Symposium, Chromosome Structure and Function: Impact of New Concepts" New York: Plenum Press.

Modi WS, Nash WG, Ferrari AC, O'Brien SJ (1987): Cytogenetic methodologies for gene mapping and comparative analyses in mammalian cell culture systems. Gene Anal Tech 4:75–85.

Morris R, Morris D (1981): "The Giant Panda" (Barzdo J, rev.). New York: Penguin Books.

Nash WG, O'Brien SJ (1987): A comparative chromosome banding analysis of the Ursidae and their relationship to other Carnivores. Cytogenet Cell Genet 45:206–212.

Nei M (1972): Genetic distance between populations. Am Nat 106:283–292.

Nei M (1978): Estimation of average heterozygosity and genetic distance from a small number of individuals. Genetics 89:583–590.

Nei M (1987): "Molecular Evolutionary Genetics." New York: Columbia Univ Press.

Newman RE, Davidson WM (1966): Comparative study of the karyotypes of several species in carnivora including the giant panda *(Ailuropoda melanoleuca).* Cytogenet 5:152–163.

O'Brien SJ (1987): The ancestry of the giant panda. Sci Am 257:102–107.

O'Brien SJ, Seuanez HN, Womack JE (1985): On the evolution of genome organization in mammals. In MacIntyre RJ (ed): "Molecular Evolutionary Genetics." Monographs Evol Biol Ser, New York: Plenum Press, pp 519–589.

O'Brien SJ, Nash WG, Wildt DE, Bush ME, Benveniste RE (1985): A molecular solution to the riddle of the giant panda's phylogeny. Nature 317:140–144.

Pei WC (1974): A brief evolutionary history of the giant panda. Acta Zool Sinica 20:188–190.

Pocock RI (1921): The external characters and classification of the Procyonidae. Proc Zool Soc London, pp 389–422.

St. Mivart G (1885): On the anatomy, classification, and distribution of the Arctoidea. Proc Zool Soc London, pp 340–404.

Sarich VM (1969): Pinniped origins and the rate of evolution of carnivore albumins. Syst Zool 18:286–295.

Sarich VM (1973): The giant panda is a bear. Nature 245:218–220.

Sarich VM, Wilson AC (1967a): Rates of albumin evolution in primates. Proc Natl Acad Sci USA 58:142–148.

Sarich VM, Wilson AC (1967b): Immunological time scale for hominoid evolution. Science 158:1200–1203.

Savage DE, Russell DE (1983): "Mammalian Paleofaunas of the World." Reading, MA: Addison-Wesley Publ Co.

Schaller GB, Jinchu H, Wenshi P, Jing Z (1985): "The Giant Pandas of Wolong." Chicago: Univ Chicago Press.

Sibley CG, Ahlquist JE (1984): The phylogeny of the hominoid primates, as indicated by DNA-DNA hybridization. J Mol Evol 20:2–15.

Sibley CG, Ahlquist JE (1987): DNA hybridization evidence of hominoid phylogeny: Results from an expanded data set. J Mol Evol 26:99–121.

Sneath PHA, Sokal RR (eds) (1973): "Numerical Taxonomy: The Principles and Practice of Numerical Classification." San Francisco: W. H. Freeman and Co.

Swofford DL (1985): "Phylogenetic Analysis Using Parsimony (PAUP), Version 2.3." Champaign, IL: Illinois Nat Hist Surv.

Thenius E (1979): Zur systematischen und phylogenetischen Stellung des Bambusbaren: *Ailuropoda melanoleuca* David (Carnivora, Mammalia). Z Saugetierk 44:286–305.

Thorpe JP (1982): The molecular clock hypothesis: Biochemical evolution, genetic differentiation, and systematics. Ann Rev Ecol System 13:139–168.

Wang T-K (1974): Taxonomic status of the species, geological distribution and evolutionary history of *Ailuropoda.* Acta Zool Sinica 20:191–201.

Wayne RK, Nash WG, O'Brien SJ (1987): Chromosomal evolution of the Canidae: II. Divergence from the primitive carnivore karyotype. Cytogenet Cell Genet 44:134–141.

Wayne RK, Benveniste RE, Janczewski DN, O'Brien SJ (1989): Molecular and biochemical evolution of the carnivora. In Gittleman JL (ed): "Carnivore Behavor, Ecology and Evolution." New York: Cornell Univ Press, pp 465–494.

Wilson AC, Carlson SS, White, TJ (1977): Biochemical evolution. Annu Rev Biochem 46:573–639.

Wurster-Hill DH, Bush M (1980): The interrelationship of chromosome banding patterns in the giant panda *(Ailuropoda melanoleuca),* hybrid bear *(Ursus middendorfi* x *Thalarctos maritimus),* and other carnivores. Cytogenet Cell Genet 27:147–154.

Wurster-Hill DH, Centerwall WR (1982): The interrelationships of chromosome banding patterns in canids, mustelids, hyena, and felids. Cytogenet Cell Genet 34:178–192.

Wurster-Hill DH, Gray CW (1973): Giemsa banding patterns in the chromosomes of twelve species of cats (Felidae). Cytogenet Cell Genet 12:388–397.

Wurster-Hill DH, Gray CW (1975): The interrelationship of chromosome banding patterns in procyonids, viverrids, and felids. Cytogenet Cell Genet 15:306–331.

Zuckerkandl E, Pauling L (1962): Molecular disease, evolution, and genetic heterogeneity. In Kasha M, Pullman B (eds): "Horizons in Biochemistry." New York: Academic Press, pp 189–225.

Index